D1447959

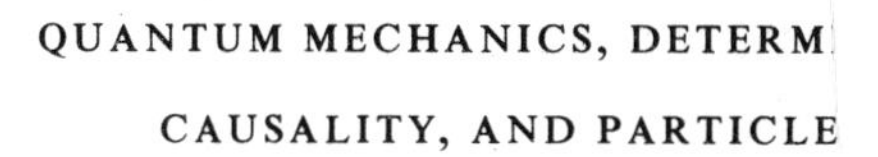

QUANTUM MECHANICS, DETERM

CAUSALITY, AND PARTICLE

MATHEMATICAL PHYSICS AND APPLIED MATHEMATICS

Editors:

M. Flato, *Université de Dijon, Dijon, France*

R. Rączka, *Institute of Nuclear Research, Warsaw, Poland*

with the collaboration of:

M. Guenin, *Institut de Physique Théorique, Geneva, Switzerland*

D. Sternheimer, *Collège de France, Paris, France*

VOLUME 1

LOUIS DE BROGLIE

QUANTUM MECHANICS, DETERMINISM, CAUSALITY, AND PARTICLES

An International Collection of Contributions in Honor of Louis de Broglie on the Occasion of the Jubilee of His Celebrated Thesis

Edited by

M. FLATO, Z. MARIC, A. MILOJEVIC, D. STERNHEIMER,
AND J. P. VIGIER

D. REIDEL PUBLISHING COMPANY
DORDRECHT-HOLLAND / BOSTON-U.S.A.

PHYSICS

Library of Congress Cataloging in Publication Data
Main entry under title:

Quantum mechanics, determinism, causality, and particles.

(Mathematical physics and applied mathematics ; v. 1)
Includes bibliographical references.
CONTENTS: Bohm, D. On the creation of a deeper insight into what may underlie quantum physical law.—Bell, J. S. The measurement theory of Everett and de Broglie's pilot wave.—Flato, M. Quantum mechanics and determinism. [etc.]
1. Quantum theory—Addresses, essays, lectures. 2. Wave mechanics—Addresses, essays, lectures. 3. Broglie, Louis, prince de, 1892– —Addresses, essays, lectures. I. Broglie, Louis, prince de 1892– II. Flato, Moshé. III. Series.
QC174.125.Q36 530.1'2 75-38666
ISBN 90-277-0623-9

Published by D. Reidel Publishing Company,
P.O. Box 17, Dordrecht, Holland

Sold and distributed in the U.S.A., Canada, and Mexico
by D. Reidel Publishing Company, Inc.
Lincoln Building, 160 Old Derby Street, Hingham
Mass. 02043, U.S.A.

All Rights Reserved
Copyright © 1976 by D. Reidel Publishing Company, Dordrecht, Holland
and copyrightholders as specified on appropriate pages
No part of the material protected by this copyright notice may be reproduced or utilized in any form or by any means, electronic or mechanical, including photocopying, recording or by any informational storage and retrieval system, without written permission from the copyright owner

Printed in The Netherlands by D. Reidel, Dordrecht

QC
174
.125
Q361
PHYS

CONTENTS

2280

PREFACE

Two important events in the history of physical sciences occurred recently: the fiftieth anniversary of Quantum Mechanics and the Jubilee of Louis de Broglie's celebrated Thesis. These events occurred in the same period of time when the world honored de Broglie on the occasion of his eightieth birthday. Some of de Broglie's friends, former students, and some people who used to know him and appreciate his personality decided to prepare an international volume for this celebrated occasion.

Such a task was not very easy. It is always simpler to contribute in honor of famous people whose works and impact were great on a technical and pragmatic level than to contribute in honor of a person whose achievements were not only dominant in physical sciences themselves, but also had many important implications for the development of the whole branch of philosophy of sciences.

Louis de Broglie, the man to whom we owe among other things the most fundamental notion of duality between waves and particles, belongs in a way to the Einsteinian school of thought. He never accepted literally the Copenhagen interpretation of quantum mechanics. To him it was clear that this interpretation makes quantum mechanics incomplete and highly non-deterministic. He always believed that since the duality between waves and particles was an experimental fact, there should be some manifestation of the Schrödinger wave itself in the realistic world. De Broglie had to struggle much for this idea, which he never gave up. The notions of hidden variables, pilot-waves, double-solution theory are all notions that we owe to him. Had we had a complete quantum-field theory, de Broglie's old dilemma of duality would have been settled, as it is clear that the structure of such a theory would contain both the field (wave) and the particle aspects.

The reader will find in this book several groups of topics. The first group of three papers treats some fundamental questions with philosophical implications, which lie at the foundations of quantum mechanics and are generally overlooked by the traditional school of physicists. We also give a short review of the historical development of hidden variables models, as well as some new ideas on the subject, and a few critical remarks concerning quantum-measurement theory and de Broglie's pilot-wave theory. These papers constitute a kind of extended introduction to the book.

The second group of three contributions deals with the question of hidden variables (already mentioned above) from the theoretical and experimental points of view. It includes a specification of Bell's original argument, a review of two recent conflicting experimental results by those who performed them, and an axiomatic study of pre-quantum models in view of a possible completion of the quantum model by the so-called hidden variables.

The two papers in the third group are, like the preceding paper, mainly concerned with the axiomatic approach to quantum theory, its relation to classical theory, and some problems of logic that are raised.

The fourth group of three papers deals mostly with some probabilistic aspects of quantum mechanics, its stochastic and statistical interpretations.

The remaining five contributions, which constitute the fifth group, treat, at various stages, the particle aspect. The first one describes a Lagrangian formalism for the Wheeler-Feynman theory of classical electrodynamics which has some intriguing properties. The second, following a line of thought of de Broglie's school and most recent theoretical developments in elementary particles theory, describes a relativistic string as a model of extended particles. The last three papers deal with de Broglie's photon theory, its extension to a description of graviton and other particles, and some possible implications of this theory in various fields.

Though we know that it is impossible to find contributions corresponding to all the variety of aspects tackled by de Broglie in his scientific activity, we have tried our utmost. We hope we have succeeded in representing some of the subjects with which de Broglie was concerned during his long and fruitful activity.

This book is dedicated to Louis de Broglie on the occasion of the jubilee of his celebrated Thesis.

November 25, 1974 THE EDITORS

D. BOHM

ON THE CREATION OF A DEEPER INSIGHT INTO WHAT MAY UNDERLIE QUANTUM PHYSICAL LAW

I should like to take this opportunity to honor Louis de Broglie on the occasion of the Jubilee by discussing something that I think has been the essential moving force within his work; i.e. a strong feeling for the necessity of inquiry into the deeper meanings of the laws of quantum physics.

I should perhaps say here that I too have always felt that quantum physics is an extremely significant clue to something deeper, and have never agreed with the common attitude, which tends to regard the quantum theory as a general form of knowledge that is final in its essence (though it may be admitted that it can perhaps still change and develop through detailed assumptions arising in particular applications). It is for this reason that I worked on 'hidden variables' for a considerable period of time. But over the years I have found that, generally speaking, my reasons for doing this work have been rather widely misunderstood. So perhaps it would be best to clear up this misunderstanding before going into the specific ideas that I would like to discuss here.

To this end, let us first consider the question: 'What is a theory'? Some light is thrown on this question by looking at the origin of the word, which is the Greek 'theoria', based on the same root as 'theatre', in a verb meaning 'to view' or 'to make a spectacle'. This would suggest that we might regard theory primarily as a way of looking at things, i.e. as a form of *insight*, rather than as 'well defined knowledge of what things are'.

To show that such a notion of theory fits in with what scientists actually do with their theories, we may consider Newton's discovery of the law of gravitation as a relevant example. Now, in ancient times, men thought that celestial matter and earthly matter were of essentially different natures, so that while it was regarded as natural for earthly objects to fall, it was also thought to be natural for celestial objects, such as the Moon, to remain up in the sky. Over the centuries, however, there arose the notion that celestial matter and earthly matter are not essentially different. But for a long time, men did not notice that this should

M. Flato et al. (eds.), Quantum Mechanics, Determinism, Causality, and Particles, 1–10. *All Rights Reserved.*
Copyright © 1976 *by D. Reidel Publishing Company, Dordrecht-Holland.*

lead to a very serious question: 'Why doesn't the Moon fall'? In a sudden flash of understanding or insight (which is of the nature of perception through the mind rather than a process of logical thought) Newton saw that the Moon *is* falling. As the apple falls toward the centre of the Earth, so does the Moon, and so indeed does all matter.

The reason why the Moon doesn't ever touch the Earth is, of course, that its tangential motion is always carrying it away from the centre, at the same time. So what is implied is a universal force of gravitation attracting all matter toward various centres, whose net result depends, however, on the particular motions of each body. And thus, one had come to a radically new way of looking at all matter, heavenly and earthly, which proved to be in harmony with general scientific experience over several centuries.

To be sure, many attempts were made to put this mode of perception on 'solid foundations' by means of precisely defined *hypotheses* from which secure conclusions could be drawn. (Indeed, as the Greek root indicates, a 'hypo-thesis' is a supposition, 'put under' the facts, which might perhaps serve as such a foundation, after it has been thoroughly tested). In my view, however, the essence of what the theorist does is in the creative act of insight, and not in the detailed hypotheses that may follow. These latter are mainly provisional and tentative suggestions for defining our forms of thought more precisely, which help to make the insight easier to relate to experiment (rather than a means of reaching a 'solid ground of knowledge' from which well defined and secure conclusions can be drawn).

As is well known, the general form of insight underlying classical physics as a whole eventually broke down, giving rise to a deep and pervasive kind of confusion or unclarity. It is important to emphasize here, however, that the broad overall hypotheses in terms of which classical physics was expressed were never actually definitively disproved by experiment. For example, it is still possible in principle to hold on to the ether hypothesis, by modifying the detailed assumptions concerning the properties of the ether in such a way that actual clocks and rulers will be inferred to obey the Lorentz transformation, as a consequence of these assumptions. However, at a certain stage, Einstein *saw* that in the situation prevailing at that time, the attempt to base everything on the ether theory was leading to serious lack of clarity. And so he dropped the old point of view and started instead from the insight that the laws of physics are to be expressed through *invariant relationships* in the actually observable phenomena and not through assumptions about the properties of a purely hypothetical ether. In a similar way, Bohr, de Broglie, Heisenberg, Schrödinger, Dirac, and others dropped old points of view and came upon equally revolutionary new insights into the laws of quantum physics.

Later, all these relativistic and quantum laws were given a 'solid foundation' through precisely formulated mathematical hypotheses. It is here, I feel, that physicists began to slip back into the old attitude that theories are to be regarded, not mainly as forms of insight, but rather as 'solid and true knowledge of the nature of reality as a whole': (the main new point of modern physics being to suppose that reality as a whole is 'relativistic and quantum mechanical' rather than 'classical' in its nature).

What I would like to suggest instead is that reality as a whole is immense and immeasurable, beyond anything that can be known as a closed and finally determined totality. Each theory is then only a *particular* form of insight, which can be regarded as a sort of light shed on certain limited aspects of reality, penetrating some limited way into the open and unknown totality. So we may expect the unending development of radically new forms of insight, rather than a steady approach to some fixed and ultimate knowledge of 'what the whole universe really is'.

Certainly, our actual experience with the development of science over the past few thousand years fits in very well with this notion of unending creation of new forms of insight, in which each form harmonizes with the content of the actual fact only within certain limits. Beyond such limits, any given form is seen to imply a kind of confusion or unclarity, which can however serve as a key clue to new forms of insight if its significance is understood in this light. And so we see the need to look carefully at unclear features of every theory, not merely trying to 'clear them up', but also keeping in mind the possibility that if they do not sooner or later clear up, they may turn out to be important indications for new forms of insight.

Most of my work has at least implicitly been in line with such a view of the nature of theory; but because the general attitude in science is to suppose that theories are mainly forms of knowledge, actual or proposed, what I have written has not conveyed its meaning properly.

For example, when I first taught the quantum theory, I found that it was very unclear; and so, I wrote a book [1] which was mainly intended as an attempt to obtain a clear insight into Bohr's views on the subject. After the book was finished, I looked at it again very carefully, and could not escape the feeling that the whole matter was still very unclear. I then began to look at it from other points of view, not to develop 'conclusive new theories about what the whole universe really is', but rather, to inquire and to see if one couldn't come upon some further insight into quantum physics. In the course of exploring several such points of view, I hit upon the notion [2] that the electron might be both a particle *and* a wave, the particle being acted on by a new kind of 'quantum potential' which was determined in a certain way by the wave intensity. I later learned that others had tried this idea before, and that indeed, de Broglie had gone into it quite carefully. De Broglie himself informed me in a letter that he had dropped the idea because it failed to meet certain objections made by Pauli [2], based on the unclear implications of the theory concerning the behaviour of scattered electrons in the two-body problem. This latter led me to look into this question more deeply, and as a result I saw that one had to extend the idea consistently to the many-body system. But in doing this, I was led to see what I felt to be a key new feature of a quantum mechanical system; i.e. that the 'quantum potential' now implied a 'many-body interaction' such that the force acting on any particle is a function of the state of the system as a *whole* as well as of the positions of all the particles in the system. So one could say that as it was necessary in Newton's time to question the generally implicit and unconsciously held notion that Heavenly objects do not fall, so the facts now available imply the need to question the notion that the behaviour of a system can be understood by analysis into parts, which interact through 'forces' that are fixed in nature, and not dependent on the state of the system as a whole. In other words, one has to look at the whole world in a new way, which is as different from that perceived by Newton, as that of Newton was from the views generally accepted earlier.

At this stage, however, one is not yet in a position to propose well defined *hypotheses* (which would perhaps correspond to that of the inverse square law of forces in classical physics). But, as pointed out earlier, the prevailing attitude in physics is that the essence of a theory is in its precisely and mathematically defined hypotheses which can be tested by experiment, and not in the insights, which latter are indeed generally considered to be useful mainly as aids to the suggestion of such hypotheses. So it was perhaps inevitable that what I wrote would be misunderstood, by being construed as a proposed hypothesis. As a hypothesis, what I was suggesting was evidently quite inadequate. But I felt that it was nevertheless a significant insight and that, as pointed out earlier, the development of such insight is the essence of the activity involved in the making of theories, rather than a mere step in coming to well defined and testable hypotheses.

The above-described insight into the possible deeper meaning of the quantum theory did indeed lead me to try various hypotheses, to give it a more detailed expression. Perhaps the most developed of these appeared in a paper that I wrote with Bub [3], which I felt brought out the key new feature implied in the quantum theory rather more clearly than had been done earlier. However, even this was quite inadequate, when considered as a specific hypothesis. But careful consideration of the whole situation then led me to see that what was involved was, more deeply, a *new notion of order*. That is, as one had to question the ancient notion of a key order between the Celestial and the Earthly, one has to question the modern Cartesian notion, that all physically relevant order is to be expressed through *systems of co-ordinates* [4]. In other words, the Cartesian co-ordinates are primarily *forms of insight*, and it is possible that they, too, are being extended beyond the context in which their meanings are free of confusion.

Through raising such questions, I came upon quite a different notion of order, which is in harmony with the general context of relativistic and quantum mechanical experience. It is not the place here, however, to describe this notion of *enfolded order* or *implicative order*. This is discussed in other articles [4, 5] in which it is shown that such a notion of order gives insight into the irrelevance of analysis of a quantum system into parts, and into the need to look on it as an undivided whole.

Meanwhile, some of the implications of this general question were taken up by other workers in the field. A very notable step was taken by Bell [6] who, on the basis of certain mathematical hypotheses, proposed a way of testing experimentally the implications of the quantum theory concerning the irrelevance of analysis of a system into 'localizable' parts (i.e. into 'elements' whose basic responses to each other did not imply an instantaneous and total interconnection or even interpenetration of all the component 'parts'). Since then, Clauser [7] and others have carried the work further, to the point at which an experimental test was indicated in detail. And this test has actually been made by Freedman and Clauser [8]. The result definitely fits in with the notion that the quantum system behaves as an indivisible whole, not relevantly analyzable into independent parts, each of which would exist in a separate region of space. And so, the need for new forms of insight into this question is made even sharper and clearer than it was before.

Another approach to this question can be obtained by considering some work done by Vigier and myself, in which we looked on the statistical features of a quantum system as originating in perturbations from a random background of motions at a deeper 'sub quantum-mechanical level' [9]. Such ideas are perhaps somewhat reminiscent of ether theories, in so far as we regard this background as universal and present everywhere in space. Indeed, modern quantum mechanical field theories imply 'zero point fluctuations' of the fields in each region of space, which are also somewhat similar, as can be seen from the fact that the properties of *all* matter are now determined in terms of quantum-mechanical averages of the 'vacuum state'; i.e. the state of so-called empty space (which is, of course, in some sense all-pervasive and omnipresent). So, in various roundabout ways, something very much like an ether theory is creeping back into physics, though, of course, the detailed properties of the new 'ether' are radically different from those considered in the nineteenth century. Nevertheless, what is most significant about this trend is perhaps that it shows in yet another context the impossibility of disproving or falsifying a general form of thought, such as the notion of a universal and basic substance or 'ether' pervading the whole of space. Indeed, it shows that even after such a form of thought has once been dropped because it leads to general confusion, it may come back again later in new ways.

Lately, de Broglie [10] has had what I regard as a very significant insight into this aspect of the question. This had led him to suggest that the deeper 'background' of space (whether we call it 'sub-quantum' or 'ether' or 'zero point fluctuations' or something else) can be regarded as a sort of 'thermal bath' with which the particle can exchange energy in the form of 'heat'. Indeed, he proposes that the 'quantum potential' originates in this exchange of 'heat energy' with the background of space.

To explain this idea, it is helpful to distinguish two kinds of motion of a system, the slow and the fast. The slow motions of a thermal system, for example, are just the ordinary motions of its parts (e.g. movement of a piston) whereby it does work and exchanges energy with other systems on the same general level as itself. The fast motions are those of its constituent molecules, whereby it exchanges energy with its environment in an 'invisible' way, i.e. a way that does not show on its own level. When such molecular motions are suitably random, this sort of exchange is called 'heat'.

But, of course, the two ways of exchanging energy are inter-related. Their relationship is given by the laws of thermodynamics. Thus, we may write

$$\mathrm{d}E = \mathrm{d}Q - \mathrm{d}W$$

where $\mathrm{d}E$ is the total change of internal energy of the system, $\mathrm{d}Q$ is the exchange through 'invisibly rapid' random motions of molecules in the form of heat, and $\mathrm{d}W$ is the exchange of energy at the 'visible' level. In general, $\mathrm{d}Q$ and $\mathrm{d}W$ are not perfect differentials, but depend on the path of change. (E.g. $\mathrm{d}W = p\,\mathrm{d}V$ for a volume, V of gas, under pressure, p). However, $\mathrm{d}E$ is always a perfect differential, and this expresses the first law of thermodynamics. The second law is expressed by saying that in thermal equilibrium, $\mathrm{d}Q = T\,\mathrm{d}S$ where T is the temperature and S the entropy.

De Broglie has in effect suggested that even the 'elementary' particles have a deeper structure, which is involved in 'fast' motions (i.e. of high frequency compared with the frequencies involved in the general motion of the particle as a whole). If this deeper structure is in 'thermal equilibrium' with the background, then there is an additional 'heat energy' involved when the particle moves from one place to another or goes from one 'quantum state' to another. And, as indicated earlier, de Broglie uses this idea to account for the 'quantum potential' acting on any given particle. By so doing, he is in effect considering the notion that the 'force' in each particle depends on parameters (such as T and S) which, in principle, imply a reference to the state of the whole universe, and which in practice can be abstracted and simplified as implying a reference to the state of a whole system of particles. So a radically new way is opened to understanding the undivided wholeness of a quantum system.

While I find this insight very illuminating, I feel that it has aspects that are not clear, and which require further examination. In particular the concept of temperature is not relativistically invariant. Because energy and momentum are components of a four-vector, any specification of temperature inevitably implies the existence of some special frame in which the average momentum is zero. But the 'vacuum state' (i.e. the state of the background when it is free of matter) should be invariant, and should not favor any particular frame. So one cannot properly attribute a temperature to the 'vacuum state'.

What this difficulty suggests to me is that energy and temperature are not appropriate concepts for describing the condition of the 'vacuum' or of 'empty space free of matter'. Rather, we need a property that is invariant. Now, the basic invariant property in the field of mechanics is action. The relevance of this property in this context is further indicated by the fact that it is just the action which is quantized.

For example, if we consider a field analyzed into normal modes of oscillation, then in the 'vacuum state' each oscillator has an average action, variable of $\mathcal{F}=h/2$. (This follows from the fact that the energy in the ground state is $E_0=h\gamma/2$ and from the relationship $\mathcal{F}=E/\gamma$). It is clear then that one can, in some rough sense, look at the 'vacuum state' as a random distribution of action among the oscillators, in which the average for each oscillator is independent of its frequency. Evidently, this is a covariant distribution. Indeed, as can easily be shown, it is invariant, not only to a Lorentz transformation, but also, to an arbitrary canonical transformation.

Let us now consider a 'particle' in a kind of 'quasi-thermal' equilibrium with its background. When considered in abstraction from this background, the action function, $F\ (X_i, t)$ of the particle will then be just the solution of the classical Hamilton-Jacobi equation, which determines a set of wave surfaces as functions of X_i and t. These equations can be obtained from the defining relation:

$$\mathrm{d}F=\sum_{i=1}^{3} p_i\,\mathrm{d}X_i-H\,\mathrm{d}t.$$

Because $\mathrm{d}F$ is, by definition, a perfect differential, we can write

$$\mathrm{d}F=\sum_{i=1}^{3}\frac{\partial F}{\partial X_i}\,\mathrm{d}X_i+\frac{\partial F}{\partial t} \tag{1}$$

and this of course yields the usual Hamilton-Jacobi equations of classical physics

$$p_\mathrm{i} = \frac{\partial F}{\partial X_\mathrm{i}} \qquad H = -\frac{\partial F}{\partial t}.$$

The new idea is then to suppose, with de Broglie, that the particle has a large number of 'fast' inner degrees of freedom,* capable of 'quasi-thermal' exchanges of action with its surroundings. Let us denote these by $\mathrm{d}A$. So Equation (1) is modified to

$$\mathrm{d}F = \sum_{i=1}^{3} p_i \,\mathrm{d}X_i - H\,\mathrm{d}t - \mathrm{d}A \tag{2}$$

$\mathrm{d}F$ is of course still to be taken as a perfect differential, defining the wave surfaces in this new context. But in a 'quasi-static' or 'equilibrium' situation we write $\mathrm{d}A = \lambda\,\mathrm{d}\mu$. Here λ is analogous to temperature and μ to entropy. (But, of course, it is *only* an analogy, because $\mathrm{d}A = \lambda\,\mathrm{d}\mu$ refers to action exchanges and $\mathrm{d}Q = T\,\mathrm{d}S$ refers to energy exchanges). We then obtain

$$\mathrm{d}F = \sum_{i=1}^{3} p_i \,\mathrm{d}X_i - H\,\mathrm{d}t - \lambda\,\mathrm{d}\mu \tag{3}$$

In the 'vacuum state', λ is rigorously a constant. But where matter is involved, λ and μ become functions of the co-ordinates, X_i and the time, t. We can thus write

$$\mathrm{d}\mu = \sum_{\mathrm{i}=1}^{3} \frac{\partial \mu}{\partial X_\mathrm{i}}\,\mathrm{d}X_\mathrm{i} + \frac{\partial \mu}{\partial t}\,\mathrm{d}t$$

and

$$\mathrm{d}F = \sum_{i=1}^{3} \left(p_i - \frac{\lambda \partial \mu}{\partial X_i}\right) \mathrm{d}X_i - \left(H - \frac{\lambda \partial \mu}{\partial t}\right) \mathrm{d}t. \tag{4}$$

The above is equivalent to adding a kind of 'vector potential' term to the Lagrangian, $L = \sum_\mathrm{i} p_\mathrm{i}(\mathrm{d}X_\mathrm{i}/\mathrm{d}t) - H$. The proposal is then to explain at least some of the characteristically new quantum properties of matter (e.g. the force that was previously attributed to the 'quantum potential') as the result of an 'environmental vector potential', which indirectly describes the connection of each particle with the whole universe through

* Cf. Heinrich Hertz, The Principles of Mechanics

'fast' random motions by which it exchanges action with the general background.

The detailed working out of this theory will be given in a later paper. For the present, I can only say that it does seem to offer the possibility of a fairly consistent description, not only of the one-particle system, but also, of the many particle system.

Finally, I would like to add that a great deal of work remains to be done before our insight will reach a point at which it will be fruitful to make well-defined and experimentally testable hypotheses involving some basically new kind of theory. What is called for now is careful attention to unclear features of the work that has been done so far and patient consideration of what is needed to clear them up.

Birkbeck College (University of London)

BIBLIOGRAPHY

[1] Bohm, D., *Quantum Theory*, Prentice Hall, N. Y., 1951.
[2] Bohm, D., *Phys. Rev.* **85** (1952), 166, 180. (See also *Causality and Chance in Modern Physics*, Routledge and Kegan Paul, London, 1957.)
[3] Bohm, D. and Bub, J., *Rev. Mod. Phys.* **38** (1966), 453.
[4] Bohm, D., *Foundations of Physics* **1** (1971), 359.
[5] 'Quantum Theory as an Indication of a New Order', *Foundations of Quantum Mechanics*, Il Corso, Academic Press N.Y., 1971.
[6] Bell, J. S., *Physics* **1** (1964), 195.
[7] Clauser, J. F., Horne, M. A., Shimony, A., and Holt, R. A., *Phys. Rev. Lett.* **23** (1969), 880.
[8] Freedman, S. J. and Clauser, J. F., *Phys. Rev. Lett.* **28** (1972), 938.
[9] Bohm, D. and Vigier, J. P., *Phys. Rev.* **96** (1954), 208.
[10] de Broglie, L., *La Thermodynamique de la particule isolél*, Gauthier-Villars, Paris, 1964. (See also L. de Broglie, *Foundations of Physics* **1** (1971), 5.)

J. S. BELL

THE MEASUREMENT THEORY OF EVERETT AND DE BROGLIE'S PILOT WAVE

In 1957 H. Everett published a paper setting out what seemed to be a radically new interpretation of quantum mechanics [1]. His approach has recently received increasing attention [2]. He did not refer to the ideas of de Broglie of thirty years before [3] nor to the intervening elaboration of those ideas by Bohm [4]. Yet it will be argued here that the elimination of arbitrary and inessential elements from Everett's theory leads back to, and throws new light on, the concepts of de Broglie [5].

Everett was motivated by the notion of a quantum theory of gravitation and cosmology. In a thoroughly quantum cosmology, a quantum mechanics of the whole world, the wave function of the world could not be interpreted in the usual way. For this usual interpretation refers only to the statistics of measurement results for an observer intervening from outside the quantum system. When that system is the whole world, there is nothing outside. This situation presents no particular difficulty for the traditional (or 'Copenhagen') philosophy, which holds that a classical conception of the macroscopic world is logically prior to the quantum conception of the microscopic. The microscopic world is described by wave functions which are determined by and have implications for macroscopic phenomena in experimental set-ups. These macroscopic phenomena are described in a perfectly classical way (in the language of 'be-ables' [6] rather than 'observables', so that there is no question of an endless chain of observers observing observers observing...). There is of course no sharply defined boundary between what is to be treated as microscopic and what as macroscopic, and this introduces a basic vagueness into fundamental physical theory. But this vagueness, because of the immense difference of scale between the atomic level where quantum concepts are essential and the macroscopic level where classical concepts are adequate, is quantitatively insignificant in any situation hitherto envisaged. So, it is quite acceptable to many people. It is not surprising then that such a consistent traditionalist as L. Rosenfeld has gone so far as to suggest [7] that a quantum theory of gravitation may be unnecessary.

M. Flato et al. (eds.), Quantum Mechanics, Determinism, Causality, and Particles, 11–17. *All Rights Reserved. Copyright © 1976 by D. Reidel Publishing Company, Dordrecht-Holland.*

The only gravitational phenomena we actually *know* are of macroscopic scale and involve very many atoms. So we only *need* the concept of gravitation on this classical level, whose separate logical status is anyway fundamental in the traditional view. Nevertheless, I think that most contemporary physicists would regard any purely classical theory of gravitation as provisional, and hold that any really adequate theory must be applicable, in principle, also on the microscopic level – even if its effects there are negligibly small[8]. Many of these same contemporary physicists are perfectly complacent about the vague division of the world into classical macroscopic and quantum microscopic inherent in contemporary (i.e., traditional) quantum theory. This mixture of concern on the one hand and complacency on the other is in my opinion less admirable than the clear headed and systematic complacency of Rosenfeld.

Everett was complacent neither about gravitation nor quantum theory. As a preliminary to a synthesis of the two he sought to interpret the notion of a wave function for the world. This world certainly contains instruments that can detect, and record macroscopically, microscopic and other phenomena. Let A be the recording part, or 'memory', of such a device, or of a collection of such devices, and let B be the rest of the world. Let the co-ordinates of A be denoted by a, and of B by b. Let $\phi_n(a)$ be a complete set of states for A. Then, one can expand the world wave function $\psi(a, b, t)$ at some time t in terms of the ϕ_n:

$$\psi(a, b, t) = \sum_n \phi_n(a)\, \chi_n(b, t) \tag{E}$$

We will refer to the norm of χ_n

$$\int \mathrm{d}b\, |\chi_n(b, t)|^2$$

as the 'weight' of ϕ_n in the expansion. As an example A might be a photographic plate that can record the passage of an ionizing particle in a pattern of blackened spots. The different patterns of blackening correspond to different states ϕ_n. Then it can be shown[9] along lines laid down long ago by Mott and Heisenberg, that the only states ϕ_n with appreciable weight are those in which the blackened spots form essentially a linear sequence, in which the blackening of neighbouring plates, or of different parts of the same plate, are consistent with one another, and so on. In the same way Everett, allowing A to be a more complicated memory, such as that of a computer (or even a human being), or a collection of such memories, shows that only those states ϕ_n have appreciable weight in which the memories agree on a more or less coherent story of the kind we have experience of. All this is neither new nor controversial. The novelty is in the emphasis on memory contents as the essential material of physics and in the interpretation which Everett proceeds to impose on the expansion E.

An exponent of the traditional view, if he allowed himself to contemplate a wave function of the world, would probably say the following. Once a macroscopic record has been formed we are concerned with fact rather than possibility, and the wave function must be adjusted to take account of this. So from time to time the wave function is "reduced"

$$\psi \to N \sum{}' \phi_n(a)\, \chi_n(b, t) \qquad \text{(E}'\text{)}$$

where (N being a renormalization factor) the restricted summation $\sum'$ is over a group of states ϕ_n which are 'macroscopically indistinguishable'. The complete set of states is divided into many such groups, and the reduction to a particular group occurs with probability proportional to its total weight

$$\sum{}' \int \mathrm{d}b\, |\chi_n|^2 .$$

He will not be able to say just when or how often this reduction should be made, but would be able to show by analyzing examples that the ambiguity is quantitatively unimportant in practice. Everett disposes of this vaguely defined suspension of the linear Schrödinger equation with the following bold proposal: it is just an illusion that the physical world makes a particular choice among the many macroscopic possibilities contained in the expansion; they are *all* realized, and no reduction of the wave function occurs. He seems to envisage the world as a multiplicity of 'branch' worlds, one corresponding to each term $\phi_n \chi_n$ in the expansion. Each observer has representatives in many branches, but the representative in any particular branch is aware only of the corresponding particular memory state ϕ_n. So he will remember a more or less continuous sequence of past 'events', just as if he were living in a more or less well defined single branch world, and have no awareness of other branches.

Everett actually goes further than this, and tries to associate each particular branch at the present time with some particular branch at any past time in a tree-like structure, in such a way that each representative of an observer has actually lived through the particular past that he remembers. In my opinion this attempt does not succeed[9] and is in any case against the spirit of Everett's emphasis on memory contents as the important thing. We have no access to the past, but only to present memories. A present memory of a correct experiment having been performed should be associated with a present memory of a correct result having been obtained. If physical theory can account for such correlations in present memories it has done enough – at least in the spirit of Everett.

Rejecting the impulse to dismiss Everett's multiple universe as science fiction, we raise here a couple of questions about it.

The first is based on this observation: there are infinitely many different expansions of type E, corresponding to the infinitely many complete sets ϕ_n. Is there then an additional multiplicity of universes corresponding to the infinitely many ways of expanding, as well as that corresponding to the infinitely many terms in each expansion? I think (I am not sure) that the answer is no, and that Everett confines his interpretation to a particular expansion. To see why suppose for a moment that A is just an instrument with two readings 1 and 2, the corresponding states being ϕ_1 and ϕ_2. Instead of expanding in ϕ_1 and ϕ_2 we could, as a mathematical possibility, instead expand in

$$\phi_{\pm} = (\phi_1 \pm \phi_2)/\sqrt{2} \quad \text{or} \quad \phi'_{\pm} = (\phi_1 \pm i\phi_2)/\sqrt{2}.$$

In each of these states the instrument reading takes no definite value, and I do not think Everett wishes to have branches of this kind in his universe. To formalize his preference let us introduce an instrument reading operator R:

$$R\phi_n = n\phi_n$$

and operators Q and P similarly related to $\phi_{\pm}$ and $\phi'_{\pm}$. Then we can say that Everett's structure is based on an expansion in which instrument readings R, rather than operators like Q or P, are diagonalized. This preference for a particular set of operators is not dictated by the mathematical structure of the wave function ψ. It is just added (only tacitly by Everett, and only if I have not misunderstood) to make the model reflect human experience. The existence of such a preferred set of variables is one of the elements in the close correspondence between Everett's theory and de Broglie's – where the positions of particles have a particular role.

The second question grows out of the first: if instrument readings are to be given such a fundamental role should we not be told more exactly what an instrument reading is, or indeed, an instrument, or a storage unit in a memory, or whatever? In dividing the world into pieces A and B Everett is indeed following an old convention of abstract quantum measurement theory, that the world does fall neatly into such pieces – instruments and systems. In my opinion this is an unfortunate convention. The real world is made of electrons and protons and so on, and as a result the boundaries of natural objects are fuzzy, and some particles in the boundary can only doubtfully be assigned to either object or environment. I think that fundamental physical theory should be so formulated that such artificial divisions are manifestly inessential. In my opinion Everett has not given such a formulation – and de Broglie has.

So we come finally to de Broglie. Long ago he faced the basic duality of quantum theory. For a single particle the mathematical wave extends over space, but the experience is particulate, like a scintillation on a screen. For a complex system, ψ extends over the whole configuration space, and over all n in expansions like (E), but experience has a particular character, like the reduced expansion (E'). De Broglie made the simple and natural suggestion: the wave function ψ is not a complete description of reality, but must be supplemented by other variables. For a single particle he adds to the wave function $\psi(\mathbf{r}, t)$ a particle co-ordinate $\mathbf{x}(t)$ – the instantaneous position of the localized particle in the extended wave. It changes with time according to

$$\dot{\mathbf{x}} = \left[Im\psi^*(\mathbf{x}, t)\frac{\partial}{\partial \mathbf{x}}\psi(\mathbf{x}, t)\right] / |\psi(\mathbf{x}, t)|^2. \tag{G}$$

In an ensemble of similar situations $\mathbf{x}$ is distributed with weight $|\psi(\mathbf{x}, t)|^2$ $d\mathbf{x}$, a situation which follows from (G) for all t if it holds at some t. To make a model of the world, a simple world consisting just of many non-relativistic particles, we have only to extend these prescriptions from 3 to $3N$ dimensions, where N is the total number of particles. In this world the many-body wave function obeys exactly a many-body Schrödinger equation. There is no 'wave function reduction', and all terms in expansions like E are retained indefinitely. Nevertheless the world has a definite configuration $(\mathbf{x}_1, \mathbf{x}_2, \mathbf{x}_3 \ldots)$ at every instant, changing according to the $3N$ dimensional version of (G). This model is like Everett's in employing a world wave function and an exact Schrödinger equation, and in superposing on this wave function an additional structure involving a preferred set of variables. The main differences seem to me to be these.

(1) Whereas Everett's special variables are the vaguely anthropocentric instrument readings, de Broglie's are related to an assumed microscopic structure of the world. The macroscopic features of direct interest to human beings, like instrument readings, can be brought out by suitably coarse-grained averaging, but the ambiguities in doing so do not enter the fundamental formulation.

(2) Whereas Everett assumes that *all* configurations of his special variables are realized at any time, each in the appropriate branch universe, the de Broglie world has a *particular* configuration. I do not myself see that anything useful is achieved by the assumed existence of the other branches of which I am not aware. But let he who finds this assumption inspiring make it; he will no doubt be able to do it just as well in terms of the $\mathbf{x}$'s as in terms of the R's.

(3) Whereas Everett makes no attempt, or only a half-hearted one, to link successive configurations of the world into continuous trajectories, de Broglie does just this in a perfectly deterministic way (G). Now these trajectories of de Broglie, innocent as (G) may look in the configuration space, are really very peculiar as regards locality in ordinary three-space[9]. But we learn from Everett that if we do not like these trajectories we can simply leave them out. We could just as well redistribute the configuration $(\mathbf{x}_1, \mathbf{x}_2, \ldots)$ at random (with weight $|\psi|^2$) from one instant to the next. For we have no access to the past, but only to memories, and these memories are just part of the instantaneous configuration of the world.

Does this final synthesis, omitting de Broglie's trajectories and Everett's other branches, make a satisfactory formulation of fundamental physical theory? Or rather would some variation of it based on a relativistic field theory? It is logically coherent, and does not need to supplement mathematical equations with vague recipes. But I do not like it. Emotionally, I would like to take more seriously the past of the world (and of myself) than this theory would permit. More professionally, I am uneasy about the possibility of incorporating relativity in a profound way. No doubt it would be possible to ensure memory of a null result for the Michelson-Morley experiment and so on. But could the basic reality be other than the state of world, or at least a memory, extended in space at a single time – defining a preferred Lorentz frame? To try to elaborate on this would only be to try to share my confusion.

CERN, Geneva

REFERENCES AND FOOTNOTES

[1] Everett, H., *Revs. Modern Phys.* **29** (1957), 454; see also Wheeler, J. A., *Revs. Modern Phys.* **29** (1957), 463.
[2] See for example:
de Witt, B. S. and others in *Physics Today* **23** (1970), No. 9, 30 and **24**, No. 4, **36** (1971) and references therein. Ideas like those of Everett have also been set out by
Cooper, L. N. and van Vechten, D., *American J. Phys.* **37** (1969), 1212 and by L. N. Cooper in his contribution to the Trieste symposium in honour of P. A. M. Dirac, September 1972.
[3] For a systematic exposition see: de Broglie, L., 'Tentative d'Interprétation Causale et Non-linéaire de la Mécani que Ondulatoire', Gauthier-Villars, Paris, 1956.
[4] Bohm, D., *Phys. Rev.* **85** (1952), 166, 180.
[5] This thesis has already been presented in my contribution to the international colloquium on issues in contemporary physics and philosophy of science, Penn. State University, September 1971, CERN TH.1424. That paper is referred to for more details of several arguments, but the opportunity has been taken here to expand on some points only mentioned there.
[6] Bell, J. S., contribution to the Trieste symposium in honour of P. A. M. Dirac, CERN TH.1582, September 1972[10].
[7] Rosenfeld, L., *Nuclear Phys.* **40** (1963), 353.
G. F. Chew has suggested that the *electromagnetic* interaction must be considered apart (although not of course left un quantized) because of its macroscopic role in observation (*High Energy Physics*, Les Houches, 1965, ed. by C. de Witt and M. Jacob, Gordon and Breach, 1965.).
[8] It is beside the present point that microscopic gravitation might not in fact be quantitatively unimportant; see, for example, the contribution of A. Salam to the Trieste symposium in honour of P. A. M. Dirac, September 1972[10].
[9] For details see the paper referred to in note 5.
[10] *The Physicist's Conception of Nature*, Ed. by J. Mehra, Dordrecht, Reidel, 1973.

M. FLATO

QUANTUM MECHANICS AND DETERMINISM

1. Prologue

During my stay with the Institute for Theoretical Physics of the Royal Institute of Technology, Stockholm, I had many occasions to discuss with people their problems concerning the foundations of Quantum Mechanics. It is needless to say that thanks to geographical conditions, most of my friends in Stockholm accepted without any further questioning the Copenhagen interpretation of quantum theory. This fact was not surprising.

What seemed surprising to me was the fact that most of them did not even bother to look at other approaches to this fundamental problem. I then decided to give a seminar lecture concerning the Einstein-Rosen-Podolsky paradox and the problem of hidden variables. I was delighted to see that not only my colleagues but also some of their younger students came to this seminar. I therefore tried to be less technical and more pedagogical. Of course, no claim of originality was associated with that seminar talk.

After having completely forgotten about that event, J. P. Vigier told me one day that Louis de Broglie's eightieth birthday was coming, and that on such an occasion some of de Broglie's friends should prepare a special issue. I was then asked to contribute to this issue and eventually to be one of its editors. After having given some thought to the problem I remembered my lecture in Stockholm and decided that this was something suitable for such an event.

Indeed, is it not Louis de Broglie who never accepted the Copenhagen interpretation of Quantum Theory? Can one deny that it was de Broglie who really pushed in the Einsteinian direction and brought others to get interested in hidden-variables theories?

This article, based on my seminar lecture in Stockholm (January 1971) is devoted to Louis de Broglie.

M. Flato et al. (eds.), Quantum Mechanics, Determinism, Causality, and Particles, 19–31. *All Rights Reserved.*
Copyright © 1976 by D. Reidel Publishing Company, Dordrecht-Holland.

2.

Let us begin by discussing the difference between *classical* (including relativistic) and *quantum* physical theories: the main difference is that in classical theories we can always divide the world into observer and observed such that the interaction between both is either negligibly small or can be controlled by us by measurements. In quantum physics this is not true! Due to the discontinuous character of changes we cannot neglect the interaction between observer and observed. It is rather that the process of observation changes the observed in an uncontrolled manner. Moreover we can give a theoretical lower bound for the accuracy of the quantum measurement in many cases. It is only by postulating such relations that ensure us such a lower bound (in case it exists) that we have a self-consistent theory of the atom – quantum mechanics.

Of course, this point of view is the traditional one, and we can still argue about a possible complexity of classical measurement theory, when such a theory exists.

The difference between the classical and the quantum cases evokes the question: Does physical reality exist by itself in the physicist's view ('objectively') when the measurement procedure is not applied? Two answers are possible:

(1) No. In this case we can accept without difficulties the usual interpretation of quantum mechanics.

(2) Yes. In this case we can believe that quantum mechanics is not complete and in a final theory, we shall be able to 'come back' to the classical concepts.

To illustrate this philosophy we present a classical paradox due to Einstein, Rosen and Podolsky.

The philosophy of these people is the following: reality is objective. Physical concepts just try to explain it and to picture it out to ourselves. A theory must answer the following criteria:

1. Correctness.
2. Completeness.

Now the first criterion is decided by experiment. The paradox of ERP deals with the second one, completeness – applied to quantum mechanics. According to ERP completeness is the fact that every element of objective physical reality has a counterpart in the theory. A kind of sufficient criterion of what they call reality is the following: If, for example, without disturbing the system we can predict with certainty the value of a physical quantity, then there is an element of the physical reality corresponding to it. Let us consider for example a particle in one dimension having the wave function $\psi=\exp((2\pi i/h)\, p_0 x)$. P being the momentum operator, $P\psi=(h/2\pi i)\,(\partial\psi/\partial x)=p_0\psi$. Here in the state ψ the momentum of the particle takes the value p_0. Therefore according to ERP it corresponds to the physical reality. On the other hand for the position (Q) of the particle we know that $Q\psi=x\psi\neq a\psi$ with constant a. Since $\psi\bar{\psi}=1$ we can just say that all values of coordinates are equally probable. Therefore the value of x is not predictable, and can only be measured directly. However when we measure it, we alter the state. Therefore in quantum mechanics we say that when momentum is known position is not predictable.

It follows that:

(1) Either the quantum-mechanical description by wave-function is not complete, or

(2) If two operators do not commute, the physical quantities they represent do not have simultaneous physical reality. (Otherwise both must be predictable by *the* wave-function).

The ERP paradox tries to prove that contrary to what people believe it is point (1) which holds.

3. The Paradox

Suppose we have two systems I and II, interacting for time $0\leqslant t\leqslant T$. Suppose that for $t>T$ there is no interaction between both parts. If we know the states of I and II for a certain $t_0>T$ we can calculate the state of I+II for instance for $t>T$. If ψ is the combined state for I+II for $t>T$, we can say something about the states of I and II for $t>T$ by the reduction technique of wave packets. Suppose $U_n(x_1)$ are eigenfunctions of an observable A in I with eigenvalues a_n. We now write $\psi(x_1, x_2)= =\sum_{n=1}^{\infty}\psi_n(x_2)\, U_n(x_1)$, $\psi_n(x_2)$ being just coefficients. If A is measured and found to have the value a_k, we conclude that the second system is now in the state $\psi_k(x_2)$ (reduction of the wave-packet). We then have $\psi(x_1, x_2)= \psi_k(x_2)\, U_k(x_1)$. Now, if we measure B (instead of A) in the first system, we conclude that the second system is in another state than before, though state of II. The same reality is here described by *two different* wave functions! In particular, suppose that the two states that we concluded that II was in – by measuring A and B in I – are eigenfunctions of two non-commuting operators. That this can be the case is illustrated by:

$$\psi(x_1, x_2) = \int_{-\infty}^{+\infty} \exp\left(\frac{2\pi i}{h}(x_1 + x_0 - x_2)\, p\right) \mathrm{d}p$$

with x_0 constant. A is the momentum of the first particle. Then $U_p(x_1) = \exp(-(2\pi i/h)\, px_1)$, U_p eigenfunction of A. We conclude that for II:

$$\psi_p(x_2) = \exp\left(-\frac{2\pi i}{h}(x_2 - x_0)\, p\right).$$

Now ψ_p is eigenfunction of the momentum of the second particle with momentum eigenvalue $-p$. If B is the coordinate of I, $V_\alpha(x_1) = \delta(x_1 - x)$, we conclude that II is in the state

$$\int_{-\infty}^{+\infty} \exp\left(\frac{2\pi i}{h}(x + x_0 - x_2)\, p\right) \mathrm{d}p = h\delta(x - x_2 + x_0)$$

which is eigenstate of the coordinate operator X_2 of II. As momentum and coordinate of II do not commute we illustrated our assertion. If this is the case, then by measuring either A or B in I (and without disturbing II) we predict with certainty either the value of P or the value of Q in II (P and Q being the non-commuting operators of before). We conclude that in the first case P is an element of reality and in the second case Q is. Both wave functions, however, belong to the same reality.

Now (conclude ERP) if quantum mechanics is *complete*, our example brings us into contradiction, because it brought us to the conclusion that two non-commuting operators can have simultaneous reality.

This contradicts the choice of 1 or 2 at the beginning. Therefore quantum mechanics is not a complete theory, Q.E.D.

It is clear that no paradox is possible if *reality* is defined in a different way (for instance the possibility of measuring simultaneously will be necessary for simultaneous reality!). However, for a classical physicist like Einstein, the fact that reality can depend on *measurement* is an idea which is strange to physical sciences.

It is also evident that much more profound an analysis can be done (and has been done) concerning the specific choice of the wave-packet reduction formula as well as around the problem of the correlations of subsystems I and II once they are 'apparently separated'.

ERP conclude their paper by a hope that a more complete theory would be possible in the future to replace quantum mechanics. And this opened the door to 'hidden variables theories' and eternal discussions between von Neumannists and Einsteinists.

4.

The arguments we presented up till now had the following aims:

(1) To show that a change in philosophy of physical reality can alter our conclusions concerning physical sciences;

(2) To give a philosophical background to the physical interpretation of uncertainty principles;

(3) To give a detailed background to the problem of hidden variables.

We therefore pass now to an introduction to the hidden variables problem. It is clear that we live in a classical world and are trying to look into the quantum mechanical world in terms of results in our classical domain. There is no problem conceptually in describing the classical world directly as it is. We can describe (e.g.) positions of material bodies $a_1, a_2, \ldots$ and many other classical parameters. A complete description of the state of the world will be given by $(a_1, a_2, a_3, \ldots, \psi)$ with both classical parameters and quantum-mechanical wave-functions.

It is an interesting question to investigate where does the boundary between classical and quantum-mechanical variables lie in our description of the state of the world. Though in practice we know when a phenomenon has to be calculated by quantum theory, this knowledge is rather approximate and is not a well established physical principle.

It is rather probable that in a final description some classical parameters will remain. On the other hand, it might be that quantum mechanics is of a provisional character, and that a further theory will eliminate quantum mechanical variables in favour of other classical variables – the so-called *hidden variables.*

Another reason for believing in the possibility of having hidden variables has to do with the statistical nature of the results we get from quantum mechanics. One can for instance imagine that random statistical fluctuations are determined by some extra hidden variables (assuming quantum mechanics is not complete).

The last reason for a possible existence of hidden variables is the existence of paradoxes of the type of that of Einstein, Rosen and Podolsky (which we saw before). Let us give a simple paradox of this kind, due to D. Bohm: we are given a pair of spin $\frac{1}{2}$ particles in the singlet ($S=0$) state, and suppose they begin to move freely in opposite directions. Suppose now that $\mathbf{t}$ is an arbitrary unit vector in $\mathbb{R}^3$. From quantum mechanics we know that if $\boldsymbol{\sigma}_1 \cdot \mathbf{t}$ yields the value $+1$, then $\boldsymbol{\sigma}_2 \cdot \mathbf{t}$ will yield the value -1. ($\boldsymbol{\sigma}_1$, $\boldsymbol{\sigma}_2$ are the Pauli spin matrices of the two particles). Now this means that no matter how the two particles are situated one relatively to the other, it is always possible to predict the spin value of one of them (in an arbitrary direction) knowing the corresponding quantity for the other particle. It seems as if some kind of hidden variables (uncontrolled by us, of course) are involved here. This prediction could even be made for an a priori local situation. Later John Bell proved that no local hidden variables can reproduce all results of quantum mechanics. But let us come to the origin of the problem of hidden variables.

A *dispersion-free state* is (roughly speaking) a state which is an eigenstate of every observable in a given family of observables. Therefore every quantum mechanical system containing at least one pair of observables which are conjugate, namely for which the uncertainty relations hold, *cannot have* any quantum dispersion-free state. Since hidden variables determine with accuracy the values of all observables, to every quantum state will correspond a family of states with *different* values of hidden variables, each member of which is dispersion-free. Therefore the search for hidden variables is equivalent in a way to the search for dispersion-free states.

It was von Neumann who utilized the fact that if in quantum mechanics an observable is a linear combination of two other observables $C=\alpha A+\beta B$, the same is true for its expectation values: $\langle C\rangle=\alpha\langle A\rangle+$ $+\beta\langle B\rangle$. Now for dispersion-free states *expectation* values coincide with eigenvalues. But eigenvalues are in general not additive! Therefore von Neumann's conclusion was that dispersion-free states (and therefore hidden variables interpretation) are not possible in quantum mechanics. To see that eigenvalues are not additive, consider the three Pauli spin matrices of a particle having spin $\frac{1}{2}$. The operator $1/\sqrt{2}\,(\sigma_x+\sigma_y)$ has the eigenvalues ± 1 and this is not equal to $1/\sqrt{2}\,(\pm 1\ \pm 1)$ which are linear combinations of eigenvalues of $1/\sqrt{2}\,\sigma_x$ and of $1/\sqrt{2}\,\sigma_y$.

Now the meaning of the non-additivity of eigenvalues is very simple: in order to measure the quantities σ_x, σ_y and $1/\sqrt{2}\,(\sigma_x+\sigma_y)$, we must have three different orientations of the Stern-Gerlach magnet. These measurements can certainly not be done simultaneously.

Therefore intuitively speaking no *a priori* additivity of eigenvalues has to be expected. It is only true that statistical averages of quantities should be additive.

The problem of hidden variables is quite subtle: on the one hand many abstract demonstrations (and quite 'general' ones) were given of their non-existence. A careful analysis in simple terms shows that all hide quite *ad hoc* 'physical assumptions'.

On the other hand it is quite trivial to construct theories containing hidden variables which do not have any physical meaning. The real problem will of course be to construct theories with hidden variables having physical meaning, capable of reproducing all known quantum-mechanical results and possessing also predictions in domains not yet covered by quantum mechanics. No example of such a theory is known to our day.

We shall now discuss an explicit model. We consider a very simple example of a spin $\frac{1}{2}$ particle in a magnetic field, originally due to L. de Broglie and D. Bohm. The Schrödinger equation is here:

$$\frac{\partial\psi(\mathbf{r},t)}{\partial t}=[-\tfrac{1}{2}\Delta+\mu(\boldsymbol{\sigma}\cdot\mathbf{H})]\cdot\psi(\mathbf{r},t).$$

The wave function ψ is here a two component Pauli spinor, $\mathbf{H}$ the magnetic field, $\boldsymbol{\sigma}$ a 3-vector built of Pauli spin matrices. We supplement the picture by a hidden (single 3-vector) variable $\boldsymbol{\lambda}$ satisfying the equation:

$$\frac{\mathrm{d}\boldsymbol{\lambda}}{\mathrm{d}t}=\frac{\mathbf{J}_\psi(\boldsymbol{\lambda},t)}{\rho_\psi(\boldsymbol{\lambda},t)},$$

where ρ is a probability-density, $\mathbf{J}$ a probability-current, defined as usual:

$$\mathbf{J}_{\psi}(\mathbf{r}, t)=\operatorname{Im}\left[\psi^*(\mathbf{r}, t)\cdot\frac{\partial}{\partial\mathbf{r}}\psi(\mathbf{r}, t)\right],\quad \rho_{\psi}(\mathbf{r}, t)=\psi^*(\mathbf{r}, t)\cdot\psi(\mathbf{r}, t).$$

(Im stands for imaginary part of and * is complex-conjugation.)

A simple calculation shows that the Schrödinger equation implies the continuity equation for our current:

$$\frac{\partial\rho_{\psi}(\mathbf{r}, t)}{\partial t}+\operatorname{div}\mathbf{J}_{\psi}(\mathbf{r}, t)=0.$$

We suppose that the quantum mechanical state ψ corresponds to an ensemble $(\boldsymbol{\lambda}, \psi)$ of (hidden) dispersion-free states, $\boldsymbol{\lambda}$ occuring with density $\rho(\boldsymbol{\lambda}, t)$ satisfying: $\rho(\boldsymbol{\lambda}, t)=\rho_{\psi}(\mathbf{r}, t)$. As a matter of fact, if the last equality holds for $t=t_0$, then in virtue of the equations of motion it will hold for all t. Remembering the equation of continuity and the equation for $\mathrm{d}\boldsymbol{\lambda}/\mathrm{d}t$, it is evident that we are going to utilize the hidden variable $\boldsymbol{\lambda}$ in our interpretation as the *real position* of the particle (known with accuracy) which replaces the quantum mechanical $\langle\psi, \mathbf{r}\psi\rangle$ – the position expectation value.

Other kinds of measurements which essentially depend on position measurements (like the measurement of a spin component in a given direction by a Stern-Gerlach experiment) will have as results *determined values* given by the aid of our hidden-variable $\boldsymbol{\lambda}(t)$.

Our scheme can be generalized to non-relativistic n-particle wave-mechanics: the state vector is here $\psi(\mathbf{r}_1,\ldots, \mathbf{r}_n, t)$ and we allow interactions between the particles. The hidden variables are n vectors $\boldsymbol{\lambda}_1,\ldots, \boldsymbol{\lambda}_n$. Equations of motion for the hidden variables are as before:

$$\frac{\mathrm{d}\boldsymbol{\lambda}_m}{\mathrm{d}t}=\frac{\mathbf{J}_m(\boldsymbol{\lambda}_1,\ldots, \boldsymbol{\lambda}_n, t)}{\rho(\boldsymbol{\lambda}_1,\ldots, \boldsymbol{\lambda}_n, t)},$$

where $\rho(\boldsymbol{\lambda}_1,\ldots, t)=|\psi(\boldsymbol{\lambda}_1,\ldots, t)|^2$ and $\mathbf{J}_m(\boldsymbol{\lambda}_1,\ldots, t)=\operatorname{Im}(\psi^*(\partial/\partial\boldsymbol{\lambda}_m)\psi)$. As before the $\boldsymbol{\lambda}$'s are distributed with probability density $|\psi|^2$, and so on. Evidently the consequences are the same as in the preceding case.

Before analyzing the advantages and the disadvantages of our scheme, one should remark that we do not discuss here what happens to the hidden variables during and after the measurement. Though the problem of measurement theory is very interesting, what we try here is to reinterpret the usual theory rather than to replace it. Of course a *meaning-*

ful hidden variables scheme (which nobody has found up till now) will in a way give direct hints on how one should formulate a coherent measurement theory.

We now list the 3 main advantages of our schemes:

(1) Though we have a 2-component theory the particle has a *determined* position $\boldsymbol{\lambda}$. The classical (hidden) picture is a particle which *does not spin.* This last fact can explain why many classical rotator pictures of the spinning electron turned out to be failures: there is no need for an exact analogy between the quantum and the hidden-variables pictures. It is just that the last one should be capable of reproducing the first one.

(2) The models just studied show that the result of a (e.g.) spin-measurement depends on the initial $\boldsymbol{\lambda}$ of the particle and on the magnetic field $\mathbf{H}$. This result *takes into account* the combination of the system and the *apparatus.*

(3) Discussing the problem of the boundary between the classical and quantum worlds we said that in any case it is reasonable to suppose that the 'final picture' will contain classical variables – probably describing 'macroscopic objects'. This is really the case in the models discussed until now.

We are now going to describe the main disadvantage of our last scheme (first observed by Bell) which in turn will force us to introduce a kind of no-go theorem for local hidden variables. The disadvantage is the following one: concentrating on a measurement at a given position $\boldsymbol{\lambda}_j$, we see that it depends also on what happens in all other 'positions' $\boldsymbol{\lambda}_1, \ldots, \boldsymbol{\lambda}_{j-1}, \boldsymbol{\lambda}_{j-1}, \ldots, \boldsymbol{\lambda}_n$. This means that the system of equations of motion for the $\boldsymbol{\lambda}_j$, though being 'local' in $\mathbb{R}^{3n}$, is '*not local*' in the physical $\mathbb{R}^3$ space. In other words this means that a measurement of one of the two ERP measuring devices can influence the response of the second (distant!) device. For people who took care to analyze the ERP paradox this might look absurd, as their explanation would have to do with an *information* propagating from the first to the second device.

The question we have to answer now, is whether this difficulty is inherent whenever we want to have 'local' hidden variables or if what we had up till now was just bad luck.

5.

To have a somehow based opinion on the question we consider a typical example due to Bell. Suppose we have a system of two spin $\frac{1}{2}$ particles prepared such that they move towards two different devices which measure spin components in directions $\hat{a}$ and $\hat{b}$. Suppose our hidden variable is λ, and $\rho(\lambda)$ is its probability distribution for the given quantum mechanical state.

The result $A(=\pm 1)$ can depend on λ and on $\hat{a}$, the result $B(=\pm 1)$ on λ and $\hat{b}$. Our notion of '*locality*' will be translated here by the fact that *A does not* depend on $\hat{b}$ neither does B depend on $\hat{a}$. The question is if e.g. the mean value of $A \cdot B$ in our hidden variable scheme can equal the quantum mechanical prediction. Now:

$$\langle A \cdot B \rangle = P(\hat{a}, \hat{b}) = \int \mathrm{d}\lambda \rho(\lambda)\, A(\hat{a}, \lambda)\, B(\hat{b}, \lambda),$$

$$\text{with} \quad A = \pm 1,\ B = \pm 1.$$

Remark: If the instruments themselves contain hidden variables in a 'local' way, we shall have to average first over instrument variables and then find the same representation as before, but this time with $\bar{A}(\hat{a}, \lambda)$, $\bar{B}(\hat{b}, \lambda)$ and $|\bar{A}| \leqslant 1$, $|\bar{B}| \leqslant 1$.

Suppose $\hat{a}'$, $\hat{b}'$ are other settings of the instruments. Calculate:

$$\begin{aligned} P(\hat{a}, \hat{b}) - P(\hat{a}, \hat{b}') &= \int \mathrm{d}\lambda \rho(\lambda)\, [A(\hat{a})\, B(\hat{b}) - A(\hat{a})\, B(\hat{b}')] = \\ &= \int \mathrm{d}\lambda \rho(\lambda)\, A(\hat{a})\, B(\hat{b})\, (1 \pm A(\hat{a}')\, B(\hat{b}')) - \\ &- \int \mathrm{d}\lambda \rho(\lambda)\, A(\hat{a})\, B(\hat{b}')\, (1 \pm A(\hat{a}')\, B(\hat{b})). \end{aligned}$$

Then

$$|P(\hat{a}, \hat{b}) - P(\hat{a}, \hat{b}')| \leqslant 2 \pm (P(\hat{a}', \hat{b}') + P(\hat{a}', \hat{b}))$$

and therefore

$$|P(\hat{a}, \hat{b}) - P(\hat{a}, \hat{b}')| + |P(\hat{a}', \hat{b}') + P(\hat{a}', \hat{b}))| \leqslant 2$$

(up till now the inequality holds both for the A, B case and the $\bar{A}$, $\bar{B}$ case).

Now we suppose $\hat{a}' = \hat{b}'$ and $P(\hat{b}', \hat{b}') = -1$ (which means that we restrict ourselves to the A, B case). The inequality becomes:

$$|P(\hat{a},\hat{b})-P(\hat{a},\hat{b}')|\leqslant 1+P(\hat{b},\hat{b}').$$

Suppose now that the two particles system was in the singlet state ($S=0$). Then quantum mechanically we know what $P(\hat{a},\hat{b})$ is in this state: $P(\hat{a},\hat{b})=-\langle\boldsymbol{\sigma}\hat{a}\cdot\boldsymbol{\sigma}\hat{b}\rangle=-(\hat{a},\hat{b})$. (Therefore in our case $P(\hat{b}',\hat{b}')=-$ $-(\hat{b}'\cdot\hat{b}')=-1$).

Thus we should also have had (if the local hidden variables picture was correct): $|-(\hat{a}\cdot\hat{b})+(\hat{a}\cdot\hat{b}')|\leqslant 1-(\hat{b},\hat{b}')$ for all unit vectors $\hat{a}$, $\hat{b}$ and $\hat{b}'$. But this does not always hold: take for example

$$\hat{a}=\left(\frac{1}{\sqrt{2}},\,0,\,\frac{1}{\sqrt{2}}\right),\quad \hat{b}=(0,0,-1)\quad\text{and}\quad \hat{b}'=(1,0,0).$$

Therefore in our example the quantum mechanical result cannot be reproduced by the aid of *local* hidden variables.

We finish this example with two remarks:

(1) In order that this example would be taken seriously from the physical point of view, it should be possible to deduce the no-go consequence from general fundamental principles without making use of so many particular assumptions.

Moreover if this example is really physically meaningful, one should be able to have the same consequences under ‘small perturbations’ of the ‘idealistic’ notion of *local* hidden variables. Indeed there exists a particular sense in which the idealistic notion of locality is not stable under perturbations – and this by itself is a draw-back to the physical meaning of the notion of local hidden variables.

In addition before comparing local-hidden and quantum predictions one has to analyze the important related question of correlation-length of the separated particles in realistic experiments.

(2) One can of course check experimentally if it is the quantum prediction which is correct, or the hidden-variable prediction. This point of view – taken by some people quite seriously – seems naive to me. Quantum mechanics was proved up till now to be very successful (whenever applicable). Hidden variables (if they exist) should be physically meaningful, able to reproduce all predictions of quantum mechanics, and have extra predictions in domains in which quantum mechanics cannot solve ‘everything’ (mildly speaking).

For completeness we present a classical, simple (maybe too simple!) example of hidden variables due to Bell. Consider a spin $\frac{1}{2}$ particle without translational motion. A quantum mechanical state is here a two-component spinor ψ, the observables being represented by $\alpha I + \boldsymbol{\beta}\cdot\boldsymbol{\sigma}$ with $\alpha \in \mathbb{R}$, $\boldsymbol{\beta} \in \mathbb{R}^3$, I the 2×2 identity matrix, $\boldsymbol{\sigma}$ the 'vector' composed of the three 2×2 Pauli spin matrices.

Evidently (by diagonalization) the measurement of such observables yields one of the eigenvalues $\alpha \pm |\boldsymbol{\beta}|$, with relative probabilities calculated from $\langle\psi, (\alpha I + \boldsymbol{\beta}\cdot\boldsymbol{\sigma})\,\psi\rangle$. We introduce now a real parameter $-\frac{1}{2} \leqslant \lambda \leqslant \frac{1}{2}$ and denote the dispersion free states by (ψ, λ). By rotation of coordinates, ψ can be brought to the form $\begin{pmatrix}1\\0\end{pmatrix}$, and we suppose that indeed ψ has this form. Denote $\boldsymbol{\beta} = (\beta_x, \beta_y, \beta_z)$. Now suppose that on the dispersion free state (ψ, λ) the measurement of $\alpha I + \boldsymbol{\beta}\cdot\boldsymbol{\sigma}$ gives *with certainty* the eigenvalue $\alpha + |\boldsymbol{\beta}|\ \mathrm{Sign}(\lambda|\boldsymbol{\beta}| + \frac{1}{2}|\beta_z|)\ \mathrm{Sign}\, X$, where $X = \beta_z$ if $\beta_z \neq 0$, $X = \beta_x$ if $\beta_z = 0$ and $\beta_x \neq 0$, $X = \beta_y$ if $\beta_z = 0$ and $\beta_x = 0$, $\mathrm{Sign}\, X = +1$ if $X \geqslant 0$ and $\mathrm{Sign}\, X = -1$ if $X < 0$.

Now quantum mechanically we know that

$$(1, 0)\,(\alpha I + \boldsymbol{\beta}\cdot\boldsymbol{\sigma})\begin{pmatrix}1\\0\end{pmatrix} = \alpha + \beta_z.$$

A simple calculation shows that this is also true while averaging on the parameter λ:

$$\int_{-1/2}^{+1/2} \mathrm{d}\lambda\,[\alpha + |\boldsymbol{\beta}|\cdot \mathrm{Sign}(\lambda|\boldsymbol{\beta}| + \tfrac{1}{2}|\beta_z|)\ \mathrm{Sign}\, X] = \alpha + \beta_z.$$

Our example – though it is impossible to give here any direct physical meaning to the hidden variable – is such that hidden variables *reproduce* the quantum-mechanical predictions, and that every dispersion-free state has *with accuracy* one of the eigenvalues $\alpha + |\boldsymbol{\beta}|$ or $\alpha - |\boldsymbol{\beta}|$.

We end our article by mentioning the following four remarks:

(1) Bell's typical example, proving the clash between quantum-mechanics and the local hidden-variables picture, has been made more concrete in a note due to E. P. Wigner.

(2) *Ad hoc* assumptions similar to those introduced by von Neumann in the famous hidden-variables theorem were also supposed in more

general demonstrations of this theorem in the framework of axiomatic quantum mechanics (Jauch, Piron, Misra, etc....).

(3) Other possible roles (besides a dynamical role and the role to 'complete' quantum mechanics) can in principle be played by some types of hidden variables. We did not discuss this matter in the present note.

(4) Since quantum fields commute in the Euclidean region, the recently developed Euclidean quantum field theory could serve as a possible framework to a less naive hidden-variables model.

To sum up, though personally I do not believe that it is the hidden variables direction that will in the future bring the big break-through to physics, I still think it is much wiser to look for more physical models than for 'no-go' theorems. The reason is that our experience in Science shows that 'no-go' theorems are always based on assumptions which – even in convincing cases like that of Bell's theorem – might still be not realized by Nature itself.

Physique Mathématique,
Université de Dijon, and Collège de France, Paris.

EUGENE P. WIGNER

ON HIDDEN VARIABLES AND QUANTUM MECHANICAL PROBABILITIES*

ABSTRACT. An argument, due originally to J. S. Bell, is somewhat simplified and made more specific. It deals primarily with a quantum mechanical system consisting of the spins of two spin-$\frac{1}{2}$ particles. It shows that a description of the quantum mechanical measurement of the spin components of these two particles by means of hidden parameters is impossible if we assume that the parameters determining the outcome of the measurement of the spin of each particle are independent of the direction in which we decide to measure the spin of the other particle. The mathematical reason for the impossibility is analyzed.

1. INTRODUCTION

It has often been suggested that the stochastic nature of the quantum mechanical measurement process is not the result of the failure of determinism. Rather, it is suggested, our inability to predict the outcomes of quantum mechanical measurements is due to the lack of knowledge of the values which some 'hidden parameters' are assuming. The values of these hidden parameters (the exact nature of which remains unspecified) do uniquely determine the behavior of the system they describe, even insofar as measurements on the system are concerned. However, the values of the hidden variables cannot be obtained directly. The quantum mechanical state vectors correspond to statistical distributions of these variables, not to definite values of them. It is for this reason that they do not suffice to determine the outcomes of quantum mechanical measurements. To be sure, the outcome of a measurement on the system narrows the range which the hidden variables may have assumed before the measurement was undertaken, and hence also sharpens their distribution after the measurement. The distribution remains sufficiently unsharp, nevertheless, so that the outcomes of some, in fact most, measurements are still unforeseeable.

The preceding paragraph described the theory of hidden variables. Objections to this theory were raised by many theorists. Von Neumann,[1] in particular, pointed to the unreasonably large variety of hidden variables which must be assumed if one wishes to account for the postulate (implicit in quantum mechanical theory) that no matter how many suc-

M. Flato et al. (eds.), Quantum Mechanics, Determinism, Causality, and Particles, 33–41. *All Rights Reserved.*
Copyright © 1976 *by The American Journal of Physics.*

cessive measurements we undertake on a system, the distribution of the hidden variables remains sufficiently unsharp so that the outcomes of measurements are as unpredictable as they were to begin with. Von Neumann's arguments have been much sharpened by others.[2] The present article, however, is based on an observation of Bell, which is different from Von Neumann's though it leads to the same conclusion. The purpose of the article is to give Bell's observation a simpler, or at least a more concrete, form.

It is rather obvious that, given any quantum mechanical measurement represented by the operator Q, one can introduce a 'hidden variable', to be denoted by q, so that the statistical distribution of this hidden variable reproduces the probabilities for the various possible outcomes $\lambda_1, \lambda_2, \ldots$ of the measurement of Q. In order to do this, it is only necessary to associate domains $D_1, D_2, \ldots$ of q with the possible measurement results $\lambda_1, \lambda_2, \ldots$ and to postulate that the distribution function $P_\psi(q)$ which corresponds to the state ψ attributes a probability to the domain D_ν which is equal to the probability that the measurement of Q on ψ yields the value λ_ν.

Furthermore, if one wishes to reproduce, by means of hidden variables, the probabilities for the outcomes of several quantum mechanical measurements, represented by operators $Q_1, Q_2, \ldots$ one can do this by introducing a hidden variable q_n for each of these measurements. One can then postulate that the outcome of the measurement of Q_n depends only on the value of q_n: one associates with each of the possible outcomes λ_ν^n of the measurement of Q_n a domain D_ν^n of the variable q_n and postulates that the measurement of Q_n yields the result λ_ν^n if q_n is in the domain D_ν^n. It yields this result no matter what the values of the other variables q are. The distribution function P which is then associated with the state vector ψ is

$$P_\psi(q_1, q_2, \ldots) = P_\psi^1(q_1)\, P_\psi^2(q_2)\, P_\psi^3(q_3) \ldots, \tag{1}$$

where $P_\psi^n(q_n)$ is the distribution function which was associated, in the preceding paragraph, with the state vector ψ in such a way that it reproduces the probabilities of the possible outcomes of the measurement of Q_n. Clearly, the definition (1) of P_ψ contains a great deal of arbitrariness. The choice of the domains D^n is arbitrary and could, in fact, depend on all the variables q_m where $m \neq n$. Also, at least as long as the spectra of the operators Q are discrete, the number of hidden variables $q_1, q_2, \ldots$ could be greatly reduced, in fact, it could be reduced to 1.

The preceding discussion considers only single measurements, i.e., does not consider successions of observations on a system. However, the outcomes of such successions of measurements, as considered for instance by Von Neumann,[1] can also be accounted for by hidden variables. In order to do this, one introduces for each succession of observations which one wishes to account for, as many hidden variables as are observations in the succession. Furthermore, the probabilities for the various values of these hidden variables are not independent of each other; there must be statistical correlations between them. Naturally, the number of hidden variables increases enormously when one does this. It is hardly necessary to give explicit formulas, analogous to (1); they can be obtained easily.

2. Bell's observation[3]

Because of the enormous amount of arbitrariness in the association of P_ψ with ψ, it is very surprising that an apparently very small (and very natural) restriction on the nature of the hidden variables renders it impossible to define a distribution P_ψ which gives for certain measurements (actually nine measurements) the same probabilities as follow from quantum mechanical theory.

The system considered by Bell consists of two particles, both with spin $\frac{1}{2}$; the measurements are the components of these spins in definite directions. There are three such directions, $\omega_1, \omega_2, \omega_3$, and the nine measurements which are considered concern: the simultaneous measurements of the two spins, the component of one in the ω_i, of the other in the ω_k direction. Since the spin component of one spin-$\frac{1}{2}$ particle in a definite direction can assume only two values, $+\frac{1}{2}$, and $-\frac{1}{2}$ (to be abbreviated subsequently by $+$ and $-$), each of the nine measurements can yield four results: the two components can be both $+$, both $-$, or the first $+$ and the second $-$, or the other way around. The λ of the preceding section can each assume only four values, corresponding to these four outcomes of the measurements. If we introduce nine variables $q_1, q_2, \ldots, q_9$, the P defined in Equation (1) can reproduce any quantum mechanical probabilities for the four possible outcomes of each of these nine measurements – in fact, can reproduce any such probabilities, whether or not consistent with quantum mechanics. The intervals defined in the preceding section subdivide the nine-dimensional space of the q into 4^9 domains and the integral of P over one of these domains gives the probability for one of the four outcomes of one of the nine measurements. No contradiction can arise as long as no further postulate is introduced.

Bell does introduce, however, the postulate that the hidden variables determine the spin component of the first particle in any of the ω directions and that this component is independent of the direction in which the spin component of the second particle is *measured*. Conversely, the values of the hidden variables also determine the spin component of the second particle in any of the three directions ω_1, ω_2, ω_3, and this component is independent of the direction in which the component of the spin of the first particle is *measured*. These assumptions are very natural since the two particles may be well separated spatially so that the apparatus measuring the spin of one of them will not influence the measurement carried out on the other. Bell calls, therefore, the assumption just introduced the locality assumption. It means that even though there may be any statistical relations between the states of the two particles, their spins are not affected by the orientation of the *apparatus* used for measuring the spin component of the other. The result of this assumption is, however, that instead of the 4^9 essentially different domains of the space of the hidden variables, we have only 2^6 essentially different domains. These can be characterized by symbols $(\sigma_1, \sigma_2, \sigma_3; \tau_1, \tau_2, \tau_3)$, all σ and τ assuming two possible values: $+$ or $-$, and the σ referring to the first, the τ to the second, particle. If the hidden variables are, for instance, in the $(+ - -; - + -)$ domain, the measurement of the spin component of the first particle in the ω_1 direction will yield the value $+$ (that is, $+\frac{1}{2}$), no matter in which direction the spin of the second particle is measured; if the component in the ω_2 and ω_3 directions is measured, the result will be $-$. Similarly, the measurement of the spin component of the second particle in the ω_1 direction will give the result $-$, no matter in which direction the component of the spin of the first particle was measured. The measurement in the ω_2 direction will give the result $+$, in the ω_3 direction $-$.

A state for which the quantum mechanical probabilities of the outcomes of the nine possible measurements of the spin components cannot be reproduced, no matter what positive probabilities we attribute to the 2^6 domains $(\sigma_1, \sigma_2, \sigma_3; \tau_1, \tau_2, \tau_3)$, is the singlet state of the two spins. Let

$(\sigma_1, \sigma_2, \sigma_3; \tau_1, \tau_2, \tau_3)$ denote henceforth the probability that the hidden parameters assume, for the singlet state of the two spins, a value lying in the domain which was denoted by this symbol. In order to calculate the quantum mechanical probabilities of the various outcomes of the nine possible measurements, let us denote the angles between the three directions $\omega_1, \omega_2, \omega_3$ by $\vartheta_{12}, \vartheta_{23}, \vartheta_{31}$ (all between 0 and π). The probability that the measurement of the spin component of the first particle in the ω_i direction and the measurement of the spin component of the second particle in the ω_k direction both give a positive result (or both give a negative result) is given by $\frac{1}{2} \sin^2 \frac{1}{2}\vartheta_{ik}$. The probability that the first measurement gives a positive, the second a negative, result (or conversely) is $\frac{1}{2} \cos^2 \frac{1}{2}\vartheta_{ik}$. These expressions can be obtained by direct calculation. They can be derived also by observing that the singlet state is spherically symmetric so that the total probability of the first particle's spin being in the direction ω_i (rather than the opposite direction) is $\frac{1}{2}$. If the measurement of the first particle's ω_1 component gives a positive result, the measurement of this component of the second particle necessarily gives a negative result. Hence, the measurement of the spin of this particle in the ω_2 direction gives a positive result with the probability $\cos^2 \frac{1}{2}\vartheta$, where ϑ is the angle between the $-\omega_1$ and the ω_2 direction. The total probability for the positive outcome of both measurements is then $\frac{1}{2} \cos^2 \frac{1}{2}\vartheta = \frac{1}{2} \sin^2 \frac{1}{2}\vartheta_{12}$, as given before. The other probabilities can be calculated in the same fashion. However, as long as the directions ω_i satisfy certain conditions, these probabilities cannot be reproduced by hidden parameters.

In order to see this, let us observe first that the value of a symbol such as $(+, \sigma_2, \sigma_3; +, \tau_2, \tau_3) = 0$ because it represents states for which the measurement of both particles' spin components in the ω_1 direction is positive. For the singlet state specified, the probability of this is zero so that the hidden parameters cannot assume values which would give a positive spin of both particles in the ω_1 direction. The same is true for the ω_2 and ω_3 directions so that the symbols $(\sigma_1, \sigma_2, \sigma_3; \tau_1, \tau_2, \tau_3) = 0$ unless $\tau_1 = -\sigma_1$, $\tau_2 = -\sigma_2$, $\tau_3 = -\sigma_3$. This, then, leaves eight of the symbols finite.

Let us now calculate the probability that the measurement of the spin component of the first particle in the ω_1 direction, and that of the second in the ω_3 direction, is positive. This is the sum of 16 terms but only two of them are nonzero:

$$\sum_{\sigma_2\sigma_3} \sum_{\tau_1\tau_2} (+\sigma_2, \sigma_3; \tau_1, \tau_2+)=$$
$$=(++-;--+)+(+--;-++)=\tfrac{1}{2}\sin^2\tfrac{1}{2}\vartheta_{31}. \qquad (2)$$

The last line gives the quantum mechanical value of the quantity in question. However, the first term of the second line refers to states which give a positive ω_2 component of the spin of the first and also a positive ω_3 component of the spin of the second particle. This term is smaller, therefore [by the value of $(-+-;+-+)$] than $\frac{1}{2}\sin^2\vartheta_{23}$. Similarly, the second term of the second line is smaller [by the value of $(+-+;-+-)$] than the probability that the measurements of the ω_1 component of the first particle's spin, and of the ω_2 component of the second particle's spin both give positive values. It is, therefore, smaller than $\frac{1}{2}\sin^2\frac{1}{2}\vartheta_{12}$. It follows, therefore, from (1) that the theory of hidden parameters can reproduce the quantum mechanical probabilities only if the three directions ω_1, ω_2, ω_3 in which the spins are measured are so situated that

$$\tfrac{1}{2}\sin^2\tfrac{1}{2}\vartheta_{23}+\tfrac{1}{2}\sin^2\tfrac{1}{2}\vartheta_{12}\geqslant\tfrac{1}{2}\sin^2\tfrac{1}{2}\vartheta_{31}. \qquad (3)$$

This inequality[4] is most easily discussed for three coplanar $\boldsymbol{\omega}$ such that $\boldsymbol{\omega}_2$ bisects the angle between $\boldsymbol{\omega}_1$ and $\boldsymbol{\omega}_3$. In this case $\vartheta_{12}=\vartheta_{23}=\frac{1}{2}\vartheta_{31}$ and (3) becomes

$$\sin^2\tfrac{1}{2}\vartheta_{12}\geqslant\tfrac{1}{2}\sin^2\tfrac{1}{2}\vartheta_{31}=\tfrac{1}{2}4\sin^2\tfrac{1}{2}\vartheta_{12}\cos^2\tfrac{1}{2}\vartheta_{12} \qquad (4)$$

or $\cos^2\frac{1}{2}\vartheta_{12}\leqslant\frac{1}{2}$ that is $\vartheta_{12}\geqslant\frac{1}{4}\pi$, or $\vartheta_{31}\geqslant\frac{1}{2}\pi$. This is the condition also if $\boldsymbol{\omega}_2$ does not bisect the angle between $\boldsymbol{\omega}_1$ and $\boldsymbol{\omega}_3$ so that the condition (3) is violated whenever the three directions are coplanar. Clearly, (3) may be violated even if the directions are not coplanar and we shall see later that there are further conditions on the three directions $\boldsymbol{\omega}$ if the measurement of the projections of the two $\frac{1}{2}$ spins forming a singlet state is to be reproducible by hidden parameters.

3. Some mathematical remarks

Mathematically, Bell's conclusion seems surprising. We had, to begin with 64 regions in the space of the hidden variables, and only nine essentially different experiments the outcomes of which were to be reproduced by the 64 (or 63) probabilities of the regions. However, the fact that the probability of a positive result of the measurement of both spin components in the ω_1 direction, for instance, vanishes for the state in question of the two spins meant that the sum

$$\sum_{\sigma_2\sigma_3} \sum_{\tau_2\tau_3} (+\sigma_2\sigma_3; +\tau_2\tau_3)=0 \tag{5}$$

vanishes, and since all 16 terms are nonnegative, all must vanish separately. However, (2) and the other similar equations (for the simultaneous probabilities of the $\pm\omega_i$, $\pm\omega_k$ directions) can still be solved in terms of the eight quantities $(\sigma_1, \sigma_2, \sigma_3; -\sigma_1, -\sigma_2, -\sigma_3)$. In fact, one parameter remains undeterminate by all these equations. One does obtain, however,

$$(+++; ---)+(---; +++)= \\ =1-\tfrac{1}{2}(\sin^2\tfrac{1}{2}\vartheta_{12}+\sin^2\tfrac{1}{2}\vartheta_{23}+\sin^2\tfrac{1}{2}\vartheta_{31}) \tag{6}$$

so that the solution in terms of the $(\sigma_1, \sigma_2, \sigma_3; -\sigma_1, -\sigma_2, -\sigma_3)$ will entail at least one negative probability if the right side of (6) is negative. One also obtains

$$(+-+; -+-)+(-+-; +-+)= \\ =\tfrac{1}{2}(\sin^2\tfrac{1}{2}\vartheta_{12}+\sin^2\tfrac{1}{2}\vartheta_{23}-\sin^2\tfrac{1}{2}\vartheta_{31}) \tag{7}$$

and two other equations obtainable from Equation (7) by cyclic interchanges of the directions ω_i. The condition that the right side of Equation (7) be positive gives Equation (3) and there are two other inequalities obtainable from (3) by cyclic interchanges of the indices 1, 2, 3. Although not very important, it may be worth noting that if the right sides of Equation (6) and (7) and the expressions obtained from the latter by cyclic interchange of the indices are all positive, all probabilities $(\sigma_1, \sigma_2, \sigma_3; -\sigma_1, -\sigma_2, -\sigma_3)$ can be chosen to be positive by setting the two terms on the left sides of Equations (6) and (7) equal. The positive nature of the four expressions given is, therefore, the necessary and sufficient condition for the possibility to interpret the spin measurements in the ω_i directions on a singlet state in terms of hidden variables.

The conditions (3), and the conditions obtained therefrom by cyclic interchanges of the indices, have the form of triangular inequalities for three sides $\sin^2\frac{1}{2}\vartheta_{ik}$. The condition which derives from (6) gives an upper limit on the circumference of the triangle in question. Were the sides of the triangles $\sin\frac{1}{2}\vartheta_{ik}$, rather than $\sin^2\frac{1}{2}\vartheta_{ik}$, the triangular conditions would be valid for all directions ω_i. Bell[3] already noted the significance of the quadratic dependence of the probabilities on the angles between the directions of observation.

Let us comment, finally, on the role of the particular state, the singlet state of two $\frac{1}{2}$ spins, which was used in the argument presented. Let $\psi_1, \psi_2, \ldots$ and $\phi_1, \phi_2, \ldots$ be orthogonal sets of states of two systems. Bell's argument, as presented above, can then be applied to all states $\sum a_n \psi_n \phi_n$ of the composite system as long as at least two a_n are different from zero. It can be applied, in particular, to the states of object plus apparatus obtained in ideal quantum mechanical measurements. The example of the singlet state of two spins $\frac{1}{2}$ was used above because the realizability of this state and of the measurements used in the argument are difficult to question.

ACKNOWLEDGMENTS

I am much indebted to Dr. Bell and Dr. Shimony for comments on a preliminary version of the present note. In particular, the inclusion of the subject of the Introduction is due to their advice.

Princeton University

NOTES

* Reprinted from *American Journal of Physics* **38** (August 1970), No. 8.

[1] The discussion of Von Neumann, most commonly quoted, is that contained in his book *Mathematische Grundlagen der Quantenmechanik* (Springer-Verlag, Berlin, 1932) and the English translation of this, *Mathematical Foundations of Quantum Mechanics* (Princeton U.P., Princeton, N.J., 1955), Secs. IV.1 and IV.2. As an old friend of Von Neumann, and in order to preserve historical accuracy, the present writer may be permitted the observation that the proof contained in this book was not the one which was principally responsible for Von Neumann's conviction of the inadequacy of hidden variable theories. Rather, Von Neumann often discussed the measurement of the spin component of a spin-$\frac{1}{2}$ particle in various directions. Clearly, the probabilities for the two possible outcomes of a single such measurement can be easily accounted for by hidden variables (see, e.g., the rest of the present section or the more specific discussion on p. 448 of Bell's article, Note 2). However, Von Neumann felt that this is not the case for many consecutive measurements of the spin component in various different directions. The outcome of the first such measurement restricts the range of values which the hidden parameters must have had before that first measurement was undertaken. The restriction will be present also after the measurement so that the probability distribution of the hidden variables characterizing the spin will be

different for particles for which the measurement gave a positive result from that of the particles for which the measurement gave a negative result. The range of the hidden variables will be further restricted in the particles for which a second measurement of the spin component, in a different direction, also gave a positive result. A great number of consecutive measurements will select particles the hidden variables of which are all so closely alike that the spin component has, with a high probability, a definite sign in all directions. However, according to quantum mechanical theory, no such state is possible. Schrödinger raised the objection against this argument that the measurement of a spin component in one direction, while possibly specifying some hidden variables, may restore a random distribution of some other hidden variables. It is this writer's impression that Von Neumann did not accept Schrödinger's objection. His point was that the objection presupposed hidden variables in the apparatus used for the measurement. Von Neumann's argument needs to assume only two apparata, with perpendicular magnetic fields, and a succession of measurements alternating between the two apparata. Eventually, even the hidden variables of both apparata will be fixed by the outcomes of many subsequent measurements of the spin component in their respective directions so that the whole system's hidden variables will be fixed. Von Neumann did not publish this apparent refutation of Schrödinger's objection.

[2] J. M. Jauch and C. Piron, *Helv. Phys. Acta* **36** (1963), 827; S. B. Kochen and E. Specker, *J. Math. Mech.* **17** (1967), 59. D. Warrington, to appear shortly. This last paper, though based on Bell's observation (Note 3), shares with Von Neumann's argument the necessity to consider a succession of many observations. A rather complete and critical review of the earlier literature was given by J. S. Bell, *Rev. Mod. Phys.* **38** (1966), 447; objections against the articles reviewed were articulated also by D. Bohm and J. Bub, *Rev. Mod. Phys.* **38** (1966), 453.

[3] J. S. Bell, *Physics* **1** (1965), 195. A more quantitative evaluation of Bell's result, together with a proposal for an experimental test, was given by J. F. Clauser, M. A. Horne, A. Shimony and R. A. Holt, *Phys. Rev. Letters* **23** (1969), 880, and this writer is indebted to these authors for having called his attention to Bell's article. See also D. Warrington, Note 2.

[4] It was pointed out by A. Shimony that Bell's inequality easily follows from (3).

STUART J. FREEDMAN, RICHARD A. HOLT, AND
COSTAS PAPALIOLIOS

EXPERIMENTAL STATUS OF HIDDEN VARIABLE THEORIES

The founders of quantum mechanics did not all agree that the new theory was the most complete possible description of the state of a physical system. The idea that the wave function represented a set of probability amplitudes which were acausally replaced by a single outcome as a result of a 'measurement' was resisted by many, most notably de Broglie and Einstein. One of the early attempts to provide a deterministic description of the behavior of individual systems by the introduction of new dynamical variables was de Broglie's 1927 "theory of the double solution," [1] which he soon abandoned in the face of much skepticism and a seemingly convincing proof by von Neumann [2] that no deterministic extension of quantum mechanics via the introduction of hidden variables was possible. A quarter of a century later, however, he re-examined the arguments of the probabilists and found them wanting. In 1953 de Broglie wrote [3]

> Von Neumann's proof apparently forbids all interpretations of probability distributions in wave mechanics by means of a causal theory with hidden parameters. Now, the theories of the double solution and of the pilot wave, though unproved, nevertheless *exist*, and one might well wonder how their existence can be reconciled with von Neumann's theory.

Had he been writing in 1967, he could have added the work of Bohm and Bub [4] and of Wiener and Siegel [5] to the list of constructive hidden variables theories, and the theorems of Jauch and Piron [6], Gleason [7] and Kochen and Specker [8] to the mounting pile of proofs of the impossibility of hidden variables [9]. This apparent paradox was resolved by an incisive article by Bell [10], in which he examined the class of hidden variable theories ruled out by each proof and showed it to be a rather limited subset of all such theories; in particular von Neumann and Jauch and Piron required additivity of expectation values of non-commuting observables (which happens to be true in quantum mechanics but is not a physically justified requirement on an arbitrary theory), while Gleason and Kochen and Specker considered only theories in which the outcome of a measurement was independent of which com-

M. Flato et al. (eds.), Quantum Mechanics, Determinism, Causality, and Particles, 43–59. *All Rights Reserved. Copyright* © 1976 *by D. Reidel Publishing Company, Dordrecht-Holland.*

patible observables were simultaneously measured ('non-contextual theories'). Remarkably enough, at the very time when the axiomatic approach seemed to have reached a dead end, possibilities of experimental verification became apparent to several investigators, and it is their work which we shall describe in this article.

1. Test of the Bohm-Bub Theory

In addition to reproducing the statistical predictions of quantum mechanics, the Bohm-Bub hidden variable theory [4] was constructed to give a detailed causal account of the 'reduction of the wavepacket' during the measurement process. However, for the short time interval immediately following a measurement it makes predictions which differ from those of quantum mechanics, and it was this feature which led Papaliolios to undertake a simple and direct experimental test [11].

According to this theory, a two-state system is described in part by the usual quantum mechanical state vector

$$|\psi\rangle = \psi_1 |a_1\rangle + \psi_2 |a_2\rangle, \tag{1}$$

in which ψ_1 and ψ_2 are complex numbers satisfying the condition

$$|\psi_1|^2 + |\psi_2|^2 = 1, \tag{2}$$

and $|a_1\rangle$, $|a_2\rangle$ are basis states in a Hilbert space. (In the experiment they are linear polarization eigenstates for an optical photon.) The description of the state is completed by specifying

$$|\xi\rangle = \xi_1 |a_1\rangle + \xi_2 |a_2\rangle, \tag{3}$$

where ξ_1, ξ_2, the hidden variables, are complex numbers. The vector $|\xi\rangle$ transforms the same as $|\psi\rangle$ but obeys a different equation of motion. In equilibrium the hidden variables of an ensemble are distributed uniformly over the hypersphere

$$|\xi_1|^2 + |\xi_2|^2 = 1. \tag{4}$$

During a measurement, of sufficiently short duration, the hidden variables change negligibly but the dominant time evolution of the ψ_i is governed by

$$\mathrm{d}\psi_1/\mathrm{d}t = \gamma \left[|\psi_1|^2/|\xi_1|^2 - |\psi_2|^2/|\xi_2|^2 \right] |\psi_2|^2 \psi_1 \tag{5}$$

and the same with 1 replaced by 2 for $d\psi_2/dt$; γ is a positive number. These equations maintain the normalization condition (2), and they predict that if $|\psi_1|>|\xi_1|$ then after a short time (for sufficiently large γ), $|\psi_1|^2 \to 1$ and $|\psi_2|^2 \to 0$, whereas if $|\psi_1|<|\xi_1|$, $|\psi_1|^2 \to 0$ and $|\psi_2|^2 \to 1$. Thus, except for a set of measure zero ($|\psi_1|^2=|\xi_1|^2$), the quantum state evolves deterministically into either $|a_1\rangle$ or $|a_2\rangle$ (apart from a phase factor) depending on the values of the hidden variables. With the equilibrium distribution given above, the usual quantum mechanical probabilities are reproduced. Immediately after the measurement, however, those systems which have been found to be in $|a_1\rangle$, say, will no longer have the uniform distribution of hidden variables and a subsequent measurement will not yield the quantum results. Since this is not observed an additional mechanism must be invoked that causes the ξ's to relax toward the equilibrium distribution.

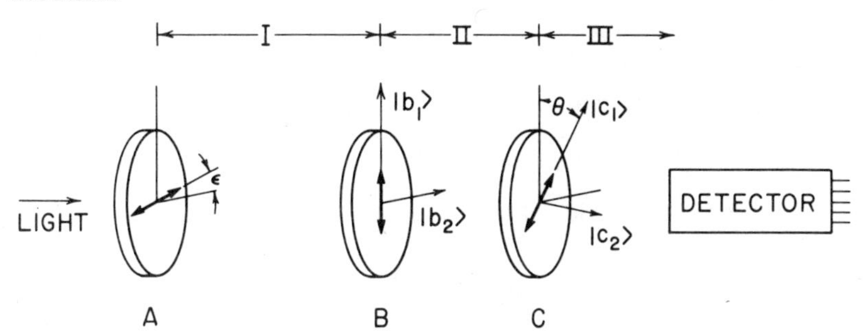

Fig. 1. Experimental arrangement for the test of the Bohm-Bub theory.

The experimental setup consisted of an incandescent light source, three polarizers, and a photomultiplier, arranged as shown in Figure 1. Photons from a low intensity lamp (to ensure that only single photons are involved in the measurements) are incident upon polarizer A, and those which are transmitted have the quantum mechanical polarization state

$$|\psi_{\mathrm{I}}\rangle = \psi_1|b_1\rangle + \psi_2|b_2\rangle = \sin\varepsilon|b_1\rangle + \cos\varepsilon|b_2\rangle \tag{6}$$

in which the $|b_i\rangle$ are basis states of linear polarization parallel and perpendicular to the axis of polarizer B, and $\pi/2-\varepsilon$ is the angle between the axes of A and B. Regardless of the distribution of the hidden variables of the photons emerging from A, the separation of the polarizers is so large that we can assume they have relaxed to the uniform distribution by the time the photons reach B. In region II, between B and C, the quantum state of the photons is

$$|\psi_{\mathrm{II}}\rangle = |b_1\rangle \tag{7}$$

and the hidden variables of those photons that emerge from polarizer B must satisfy initially

$$|\xi_1| < |\psi_1| = \sin\varepsilon \tag{8}$$

which can be a very stringent requirement on the possible values of the hidden variables if ε is made sufficiently small, though the light available to the detector is thereby reduced. This shows very clearly how polarizer B selects photons according to their usual quantum states also selects them according to their hidden variables since these variables play a role in the act of measurement.

The third polarizer C has its axis at θ relative to that of B; thus the basis states $|b_i\rangle$ can be written

$$\begin{aligned} |b_1\rangle &= \cos\theta\,|c_1\rangle - \sin\theta\,|c_2\rangle \\ |b_2\rangle &= \sin\theta\,|c_1\rangle + \cos\theta\,|c_2\rangle. \end{aligned} \tag{9}$$

The amplitude $\psi_1^{(c)} = \langle c_1 | \psi_{\mathrm{II}}\rangle$ is given by

$$\psi_1^{(c)} = \cos\theta\,\psi_1^{(b)} + \sin\theta\,\psi_2^{(b)} = \cos\theta, \tag{10}$$

while the hidden variables, which transform by the same unitary transformation, are related by

$$\xi_1^{(c)} = \cos\theta\,\xi_1^{(b)} + \sin\theta\,\xi_2^{(b)}. \tag{11}$$

In order for a photon to be transmitted by polarizer C it must have

$$|\psi_1^{(c)}| > |\xi_1^{(c)}|. \tag{12}$$

Using (4), (10) and (11) we obtain[1]

$$\frac{1-\tan^2\theta}{2\tan\theta} > \left|\frac{\xi_1^{(b)}}{\xi_2^{(b)}}\right| \quad \cos\alpha, \tag{13}$$

where α is the phase angle between $\xi_1^{(b)}$ and $\xi_2^{(b)}$. Now since

$$\left|\frac{\xi_1^{(b)}}{\xi_2^{(b)}}\right| \leqslant \tan\varepsilon, \quad \cos\alpha \leqslant 1, \tag{14}$$

a photon will certainly be transmitted if

$$\frac{1-\tan^2\theta}{2\tan\theta} > \tan\varepsilon, \tag{15}$$

or equivalently

$$\tan\left(\frac{\pi}{4}-\theta\right) \Big/ \left[1-\tan^2\left(\frac{\pi}{4}-\theta\right)\right] > \frac{\tan\varepsilon}{2}. \tag{16}$$

Hence the photon gets through if $0<\theta<\pi/4-\varepsilon/2$; a similar argument shows that the photon is definitely rejected if $\pi/4+\varepsilon/2<\theta<\pi/2$. This prediction is of course in conflict with the $\cos^2\theta$ transmission probability given by quantum mechanics (Malus' Law). In this experiment $\varepsilon=10°$, so the Bohm-Bub theory predicts certain transmission for $0<\theta<40°$ and certain rejection for $50°<\theta<90°$. The experiment verified the quantum prediction; thus the relaxation time for the hidden variables had to be shorter than the transit time for light from the front surface of B to the front surface of C. (It can be shown [11] that 90% of the photons interact in the first 3×10^{-4} cm of the Polaroid HN-32.) Since the B polarizer was only 15×10^{-4} cm thick, an upper limit of 1.9×10^{-14} [2] sec was set on the relaxation time. By equating

$$\tau \approx h/kT, \tag{17}$$

Bohm and Bub estimate $\tau\approx10^{-13}$ s for room temperature, but there is no theoretical justification for this estimate. Thus, although this experiment weakens the position of the Bohm-Bub theory it does not rule it out altogether. We note that a non-zero relaxation time τ is essential for a testable Bohm-Bub theory. As the upper limit is reduced it becomes more difficult to invent a believable physical process that might be responsible for it. The experiment has never been repeated; it is obviously worthwhile to corroborate and improve these results.

2. Test of Local Hidden Variable Theories

As Belinfante [12] has emphasized, much of the impetus for introducing hidden variables has come from physicists who were dissatisfied not merely with the indeterminism of quantum mechanics but also with its apparent non-locality. In considering measurements on spatially separated parts of a system, one must according to the Copenhagen Interpretation suppose that a measurement of one subsystem 'reduces' the state vector of the entire system. Thus, although there is in fact no violation of locality, one is left with the uneasy feeling that a measurement in one region has instantaneously changed 'something' about a far-off region, even if that something is only our *knowledge* of the state of the distant region. It was this fundamental problem that led Einstein, Podolsky, and Rosen (EPR) [13] to invent their well-known 'paradox', and which later suggested to Bell [14] an ingenious way to distinguish some of the predictions of *all* local deterministic hidden variable theories from those of quantum mechanics.

The original EPR analysis was intended to show that quantum mechanics is an incomplete theory. It has since become traditional to rephrase their discussion by introducing a version of the mathematically simpler gedankenexperiment suggested by Bohm and Aharonov [15]. Consider, for example, a system consisting of two photons moving in opposite directions (the $+z$ and $-z$ directions), for which the spin part of the wave function is [3]

$$|\psi\rangle = \frac{1}{\sqrt{2}} [|x\rangle_1 |x\rangle_2 - |y\rangle_1 |y\rangle_2] \tag{18}$$

where $|x\rangle_i$ is the eigenstate of photon i with linear polarization along the x-axis and $|x\rangle_i$ is the state along the y-axis. This type of state, which cannot be written in product form, is produced in atomic cascades and in positronium annihilation, for example.

EPR noted that if one determines that the polarization of photon 1 is along the x- or y-axis then one can predict with certainty that the other photon will be found to have a polarization along the same axis, yet clearly a measurement of photon 1 cannot physically disturb photon 2. Now the basis states in which $|\psi\rangle$ is expressed are arbitrary; it can also be written in terms of rotated x'- and y'-axes. Again a measurement of the polarization of photon 1 with respect to a primed axis allows one to predict the polarization of photon 2 along the same axis. Yet the specification of the polarization of photon 2 in both the unrotated and rotated frames is more than is allowed by quantum mechanics, since these observables do not in general commute. The conclusion drawn by EPR was that since we can predict the polarization of photon 2 with respect to any axis without disturbing it, the result of a polarization measurement in any direction must have been determined beforehand; this much information is not, however, contained in the quantum mechanical wavefunction and thus quantum mechanics is incomplete. There have been numerous replies to this argument; Bohr's [16] is probably the best known. Nevertheless it is fair to say that many physicists are still not satisfied that the questions raised by the EPR paradox have been conclusively answered.

Bell realized that EPR had done more than identify the feature of quantum mechanics most objectionable to one's intuition. He saw that the long range correlations made possible by the non-factorable form of $|\psi\rangle$ in (18) might well exceed any that were possible in a theory in which the state of each photon could be described *completely*. His re-examination of the EPR paradox led in fact to the remarkable discovery that an upper bound could be set to the strength of the correlation allowed by any deterministic hidden variable theory that satisfies a natural condition of 'locality'. In certain situations the statistical predictions of quantum mechanics violate Bell's inequality; thus it is not possible to escape from the EPR paradox by introducing hidden variables unless one is prepared to say that quantum mechanics is incorrect [17].

In order to make clear the crucial assumption of locality we describe Bell's method of proving his inequality. Polarization measurements on photon 1 are associated with a deterministic function $A(\hat{a}, \lambda)$, in which $\hat{a}$ represents the parameters of the measuring device (in this case the orientation of the polarizer) and λ represents the postulated hidden variables which give sufficient information to determine the results of *all* measurements that might be made on that photon. $A(\hat{a}, \lambda)$ takes on values corresponding to the results of a measurement: we take -1 for transmission of the photon in the ordinary ray and $+1$ in the extraordinary ray. Similarly, we associate $B(\hat{b}, \lambda)$ with photon 2. To describe an ensemble of experiments we define a measure on the hidden variables phase space represented by a positive function $\rho_{|\psi\rangle}(\lambda)$, satisfying

$$\int_{\Gamma} \rho_{|\psi\rangle}(\lambda)\, \mathrm{d}\lambda = 1, \tag{19}$$

where Γ is the hidden variable space. The subscript on ρ reminds us that this distribution will correspond to the quantum mechanical ensemble described by the state $|\psi\rangle$. To describe correlated measurements we construct the correlation function $D(\hat{a}, \hat{b})$:

$$D(\hat{a}, \hat{b}) = \int_{\Gamma} \rho_{|\psi\rangle}(\lambda)\, A(\hat{a}, \lambda)\, B(\hat{b}, \lambda)\, \mathrm{d}\lambda \tag{20}$$

which is the hidden variables analog of the quantum expectation value. The form of the correlation function given by (20) incorporates the essential assumption of locality: the function describing the product of two polarization measurements at widely separated locations has been written as the product of a function $A(\hat{a}, \lambda)$ for photon 1 which is independent of $\hat{b}$, the disposition of the measuring apparatus for photon 2, times a function $B(\hat{b}, \lambda)$ which is independent of $\hat{a}$.

With the above expression for $D(\hat{a}, \hat{b})$ and the previously defined properties of $\rho_{|\psi\rangle}(\lambda)$, $A(\hat{a}, \lambda)$ and $B(\hat{b}, \lambda)$, it can be shown [18] that the following inequality must be obeyed by the correlation function:[4]

$$-2 \leqslant D(\hat{a}, \hat{b}) - D(\hat{a}, \hat{c}) + D(\hat{d}, \hat{c}) + D(\hat{d}, \hat{b}) \leqslant 2. \tag{21}$$

The quantum mechanical expectation value is given by

$$E(\hat{a}, b) = \langle\psi|\theta_1(\hat{a})\, \theta_2(\hat{b})|\psi\rangle, \tag{22}$$

where $\theta_1(\hat{a})$ is the quantum observable for a polarization measurement of the photon in the $\hat{a}$ direction, and similarly for $\theta_2(\hat{b})$. If we take in particular the state $|\psi\rangle$ from (18) and assume perfect polarizers, then the quantum prediction is

$$E(\hat{a}, \hat{b}) = E(\phi) = \cos 2\phi \tag{23}$$

where

$$\phi = \cos^{-1}(\hat{a}\cdot\hat{b}). \tag{24}$$

Choosing the directions $\hat{a}$, $\hat{b}$, $\hat{c}$, $\hat{d}$ such that

$$\cos^{-1}(\hat{a}\cdot\hat{b}) = \cos^{-1}(\hat{d}\cdot\hat{c}) = \cos^{-1}(\hat{d}\cdot\hat{b}) = \pi/8, \tag{25}$$

$$\cos^{-1}(\hat{a}\cdot\hat{c}) = 3\pi/8,$$

we find

$$E(\hat{a}, \hat{b}) - E(\hat{a}, \hat{c}) + E(\hat{d}, \hat{c}) + E(\hat{d}, \hat{b}) = 2\sqrt{2}. \tag{26}$$

Hence the restriction by Bell's inequality, (21), makes it impossible for the hidden variables correlation to equal the quantum mechanical prediction, no matter what $\rho_{|\psi\rangle}(\lambda)$ we choose.

For the idealized correlation considered so far (and for the experiments to be discussed) it is reasonable to assume that $D(\hat{a}, \hat{b})$ depends only on the angle ϕ between $\hat{a}$ and $\hat{b}$.[5] Then using the notation $D(\phi) = D(\hat{a}, \hat{b})$ we can obtain a special consequence of inequality (21) that will be useful in the subsequent discussion. We simply note that the angles chosen in the previous example are in fact the angles for which quantum mechanics predicts a maximum violation of the righthand side of inequality (21). Furthermore, the maximum violation of the lefthand side occurs when $\cos^{-1}(\hat{a}, \cdot \hat{b}) = \cos^{-1}(\hat{d} \cdot \hat{c}) = \cos^{-1}(\hat{d} \cdot \hat{b}) = 3\pi/8$ and $\cos^{-1}(\hat{a} \cdot \hat{c}) = 9\pi/8$. Since the physical angles between the polarizer axes are equivalent modulo π, we combine the resulting inequalities, obtaining the more convenient inequality

$$|D(\pi/8) - D(3\pi/8)| \leqslant 1, \tag{26}$$

for a local hidden variable theory.

Ineq. (26) is also violated by the quantum mechanical prediction for the polarization correlation arising from the quantum state

$$|\psi\rangle = \frac{1}{\sqrt{2}}(|x\rangle_1 |y\rangle_1 + |y\rangle_2 |x\rangle_2), \tag{27}$$

for which $E(\phi) = \sin 2\phi$.

3. Experiments

The consequences of Bell's argument were analyzed from an experimental point of view by Clauser, Horne, Shimony, and Holt [18, 19] in 1969 and 1970. They concluded there was insufficient evidence that the restrictions imposed by a local hidden variable theory were, in fact, violated by nature, suggesting, however, that more conclusive experiments were possible. For a decisive experiment, the quantum predictions for the quantities actually measured must directly violate the local hidden variable restriction. Following this suggestion, experiments were carried out by Freedman and Clauser [23, 24] and by Holt and Pipkin [25, 26].

Both experiments investigate the linear polarization correlation of photons emitted from an atomic cascade. In each case decaying atoms are viewed by two symmetrically placed optical systems each containing lenses, an interference filter, a linear polarizer, and a photomultiplier capable of single photon counting. Coincidence counting techniques are used to insure that a particular two photon system originated from the same decaying atom. The experiments, then, consist of measuring coincidence count rates under various polarizer orientations (including polarizer removed).

In a real polarizer some photons are lost by absorption or scattering and thus not transmitted in either the ordinary or extraordinary ray. The argument leading to inequality (21), however, merely relies on the dichotomic nature of the measurements. We therefore interpret $+1$ (i.e. $A(\hat{a}, \lambda) = +1$ or $B(\hat{b}, \lambda) = +1$) to mean transmission of the photon and -1 to mean non-transmission, and the local hidden variable restriction will still apply to real experiments without additional assumptions.

If all photons incident on a detector were counted the correlation function would be related to the experimental coincidence rates by the equation:

$$D(\phi) = 4\frac{R(\phi)}{R_0} - 2\frac{R_1}{R_0} - 2\frac{R_2}{R_0} + 1, \tag{28}$$

where $R(\phi)$ is the coincidence rate for two photon detection with polarizers inserted at relative angle ϕ, $R_1 (R_2)$ is the coincidence rate with polarizer 1(2) inserted and 2(1) removed,[6] and R_0 is the coincidence rate with both polarizers removed. However, since the detectors used in these experiments were not perfectly efficient (the efficiencies were in all cases less than 30%) an additional assumption must be made in order to employ Equation (28). Stated more strongly than necessary, we assume that the probability of detection of a photon is independent of whether or not it has passed through a polarizer.[7]

Using Equation (28) inequality (26) can be written in the form,

$$\delta \equiv \left| \frac{R(3\pi/8)}{R_0} - \frac{R(\pi/8)}{R_0} \right| - \tfrac{1}{4} \leqslant 0. \tag{29}$$

Freedman and Clauser utilized a ${}^1S_0 \to {}^1P_1 \to {}^1S_0$ (Figure 2) cascade in atomic calcium to produce correlated two-photon systems. The calcium was formed in a beam which intersected the light from a filtered deuterium arc lamp. Calcium atoms were excited to the $3d4p\ (^1P_1)$ state by resonance absorption of light at 2275 Å. Some of the excited atoms decayed to the

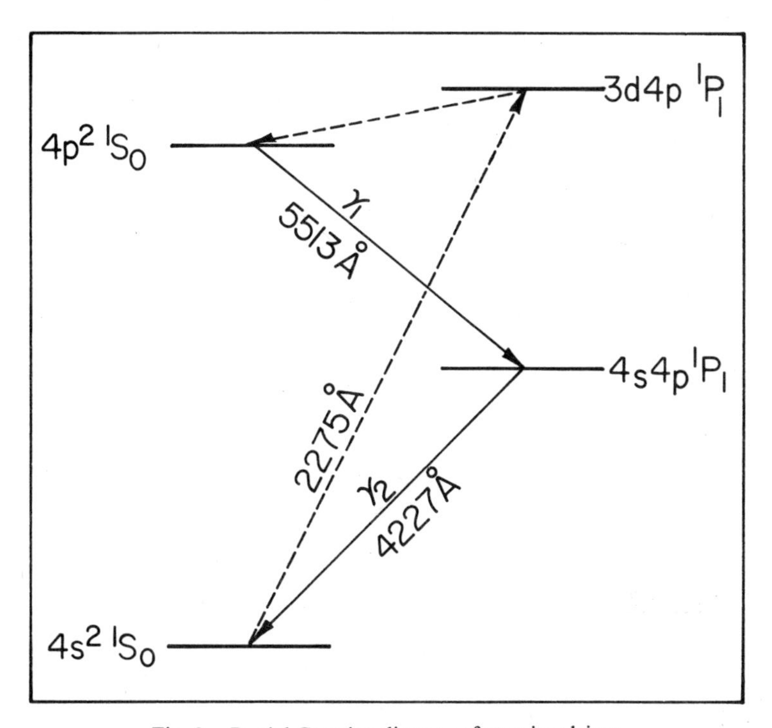

Fig. 2. Partial Grotrian diagram of atomic calcium.

$4p^2\,(^1S_0)$ state and then cascaded to the ground state with the emission of two photons at 5513 Å (γ_1) and 4227 Å (γ_2). Figure 3 shows the experimental geometry.

In the ideal case where photons γ_1 and γ_2 travel in opposite directions the two-photon state is represented in quantum mechanics by Equation (18). For this experiment quantum mechanics predicts the ratio of $R(\phi)/R_0$ to be: [18, 19, 23]

$$R(\phi)/R_0 = \tfrac{1}{4}(\varepsilon_M^1 + \varepsilon_m^1)(\varepsilon_M^2 + \varepsilon_m^2) + \\ + \tfrac{1}{4}(\varepsilon_M^1 - \varepsilon_m^1)(\varepsilon_M^2 - \varepsilon_m^2)\, F_1(\phi) \cos 2\phi, \qquad (30)$$

where $\varepsilon_M^i(\varepsilon_m^i)$ is the transmittance of the ith polarizer for light polarized parallel (perpendicular) to the polarizer axis, and $F_1(\theta)$ is a function of the half-angle θ of the detector solid angles. (This angle was the same for both detectors.) $F_1(\theta)$ represents a depolarization due to non-collinearity of the two photons, and approaches unity for infinitesimal detector solid

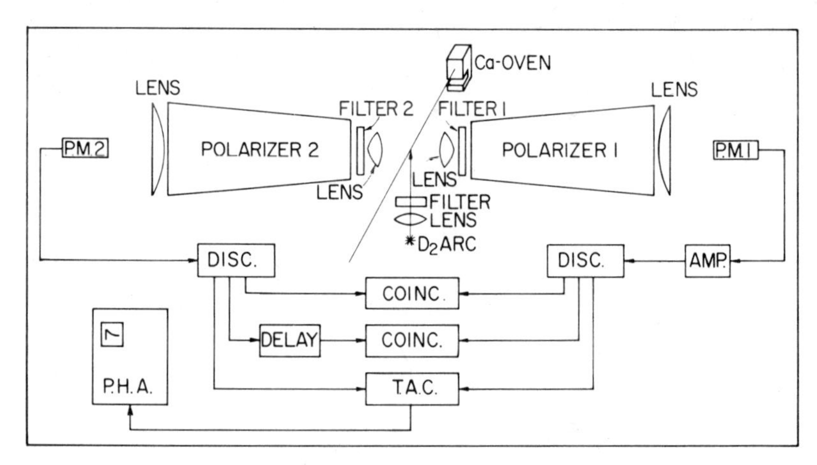

Fig. 3. The experimental arrangement for the calcium experiment. The mercury experiment is similar except for the light source and polarizers.

angles. (For this experiment $\theta \approx 30°$, and $F_1(30°) \approx .99$.)

Substituting the quantum predictions into the equation for δ (Equation (29)) we find:

$$\delta_{\text{QM}} = \frac{1}{2\sqrt{2}}(\varepsilon_M^1 - \varepsilon_m^1)(\varepsilon_M^2 - \varepsilon_m^2) F_1(\theta) - \tfrac{1}{4}. \tag{31}$$

The condition for a decisive experiment, $\delta_{\text{QM}} > 0$, is thus seen to place restrictions on the polarizer transmittances and the detector solid angles.

Using measured polarizer transmittances [23] and detector solid angles the quantum prediction for this experiment is

$$\delta_{\text{QM}} = .051. \tag{32}$$

The measured value of δ,

$$\delta = .050 \pm .008 \tag{33}$$

is in excellent agreement with the above prediction and violates inequality

(29) by more than six standard deviations. Furthermore, $R(\phi)/R_0$ was measured for eleven angles between 0° and 90° (Figure 4); these results were all consistent with quantum mechanics. Thus, this experiment provides strong evidence against local hidden variable theories.

Holt and Pipkin and [25, 26] utilized the $9\ {}^1P_1 \rightarrow 7\ {}^3S_1 \rightarrow 6\ {}^3P_0$ cascade in ^{198}Hg (Figure 5). Isotopically pure mercury was contained in a pyrex cell, and atoms were excited to the initial state of the cascade by electron excitation. The quantum state for the 5676 Å (γ_1) and the 4047 Å (γ_2)

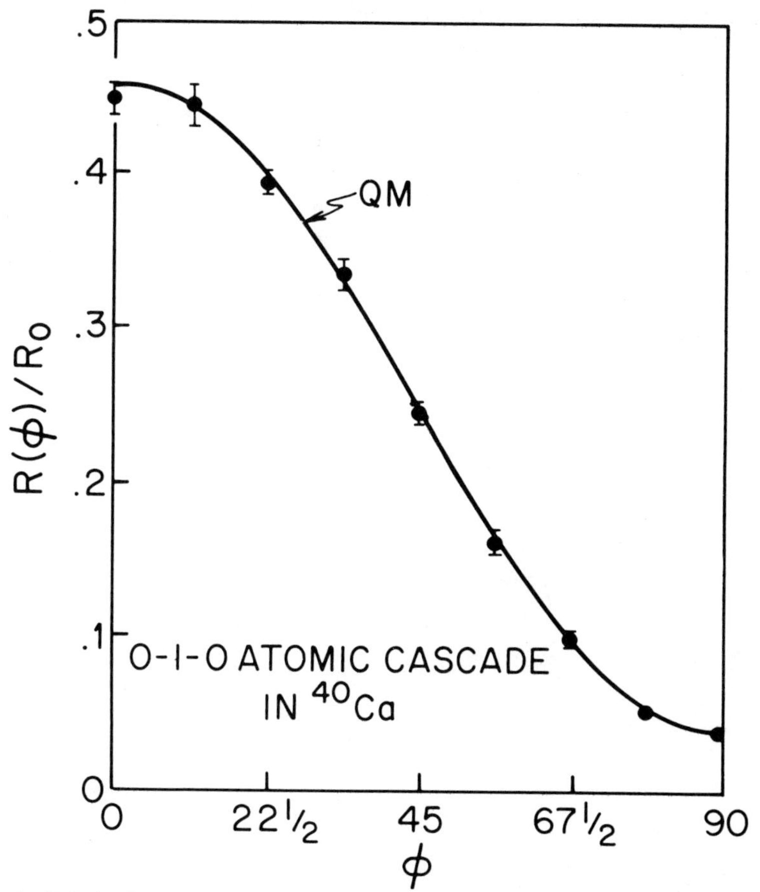

Fig. 4. Polarization correlation of photons from the calcium cascade. The solid line is predicted from Equation (30) with the measured polarizer transmittance and solid angles.

photons in the ideal case of opposite and collinear emission is given by Equation (27).

The quantum prediction for $R(\phi)/R_0$ is: [18, 19, 25]

$$\frac{R(\phi)}{R_0}=(\varepsilon_M^1+\varepsilon_m^1)(\varepsilon_M^2+\varepsilon_m^2)-\tfrac{1}{4}(\varepsilon_M^1-\varepsilon_m^1)(\varepsilon_M^2-\varepsilon_m^2)\,F_2(\theta)\cos 2\phi, \tag{34}$$

where ε_M^i and ε_m^i are defined as before but $F_2(\theta)$ is a more rapidly falling function of the detector solid angle. Using the measured polarizer transmittances [25] and experimental solid angle ($\theta\approx 13°$ and $F_2(13°)\approx .95$), the quantum mechanical prediction for δ is:

$$\delta_{QM}=+.016. \tag{35}$$

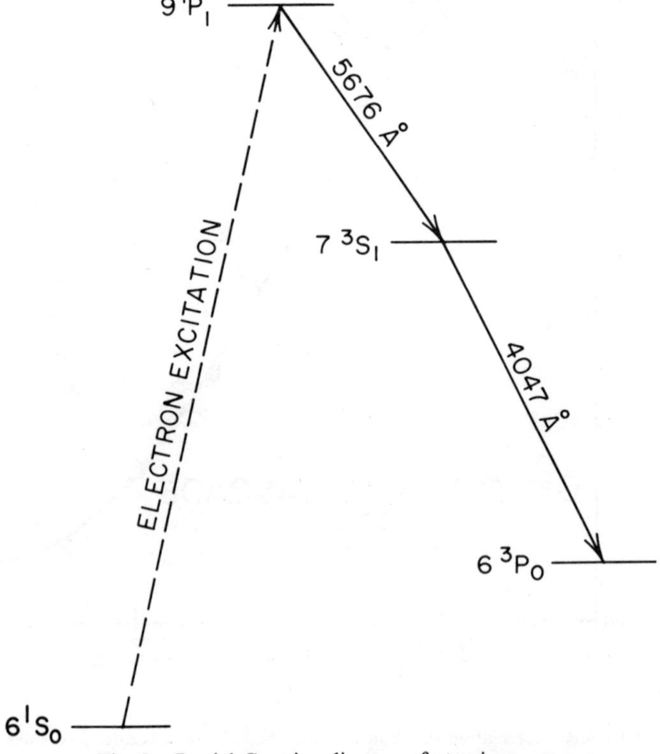

Fig. 5. Partial Grotrian diagram of atomic mercury.

However, the experimental values of $R(\pi/8)/R_0$ and $R(3\pi/8)/R_0$ yielded

$$\delta = .034 \pm .013. \tag{36}$$

This value satisfies the hidden variable restriction ($\delta \leqslant 0$) and moreover is nearly four standard deviations from the quantum mechanical prediction.

Neither Holt and Pipkin nor other experimenters have been able to find systematic errors sufficient to explain this result. However, it is generally agreed that nearly all possible systematic errors would tend to wash out the quantum mechanical polarization correlation, and Holt and Pipkin remain cautious about the correctness of their findings.

Other evidence on this question comes from experiments measuring the polarization correlation of annihilation photons from positrons stopped in matter [27–35]. It can be shown from angular momentum and parity considerations that the two-photon quantum state in the ideal case of collinear photons is given by (27). Unfortunately, the Compton polarimeters which must be used in practice are so imperfect that no test of Bell's inequality is possible [19] unless further assumptions are made about the behavior of photons in the polarizers. [34] However, agreement with quantum mechanics does somewhat weaken the position of local hidden variable theories. Most of the results, except for some early discrepancies, are in accord with quantum mechanics; however, the most recent experiment [35] seems to disagree with quantum mechanics and to lie just on the border of the hidden variable limit. Furthermore, the data vary with the distance between source and polarizers. The conclusions of this work should be contrasted with those of the comprehensive experiments of Kasday *et al.* [34] and of Langhoff [33].

We note, in conclusion, that the problem of the validity of local hidden variable theories rests with the experimentalists. New experiments are in progress or are being planned by several groups and we can hope for a solution in the near future. It is fair to say that the existing evidence still favors quantum mechanics; nevertheless, the question is of fundamental importance and there is too much at stake to allow any experimental discrepancy to remain unexplained.

Princeton University
Brown University
Harvard University

NOTES

[1] Note that Ref. [11] incorrectly has 4 in the denominator of the left-hand side.
[2] The experimental verification of Malus' law, that the intensity in region III varies as $\cos^2\theta$, was performed in steps of 10° for $0<\theta<90°$. The descrepancy between the observed curve and that predicted by Bohm-Bub is greatest at the measured angles $0=40°$, 50°. If we assume that the Bohm-Bub result at these angles decayed exponentially in time to within three times the measurement error we obtain a limit on the relaxation time τ given by

$$e^{-t/\tau}\sin^2(40°)\leqslant 3\times(0.01)$$

where τ is the time between the photon's interaction with polarizers B and C, and is 5×10^{-14} s. This results in a relaxation time $\tau\leqslant1.9\times10^{-14}$ s. This limit is smaller than the limit quoted in Ref. 11 (2.4×10^{-14} s). The reanalysis accounts for the factor of 2 error in the equation corresponding to Equation (16).
[3] Note that we are using separate right-handed coordinate systems for the two photons.
[4] This form of the restricted is more general than Bell's original. The right-hand inequality was first derived in Ref. [18]. Various proofs are described in Refs. [19–23].
[5] Although this assumption is not required by a local hidden variable theory, a theory that did not satisfy this condition would immediately disagree with quantum mechanics. In any case the assumption is experimentally testable. (Ref. [23]).
[6] R_1 and R_2 are assumed to be independent of angle. Again, a theory which did not satisfy this condition would immediately disagree with quantum mechanics and furthermore it is a testable assumption. (Ref. [23]).
[7] In fact a very special dependence is required to convert a local hidden variable correlation emerging from the polarizers into one consistent with quantum mechanics. Nevertheless, since this assumption is not testable dedicated advocates of local hidden variables could argue that these experiments are not 'completely' decisive. However, there are more serious problems at the level of present experiments.

BIBLIOGRAPHY

[1] de Broglie, L., *C.R. Acad. Sci. Paris* **183** (1926), 447, **184** (1927), 273; *J. Phys. Rad.* (6) **8** (1927), 225.
[2] von Neumann, J., *Mathematische Grundlagen der Quantenmechanik*, J. Springer, Berlin, 1932 (Eng. tr. *The Mathematical Foundations of Quantum Mechanics*, Princeton Univ. Press, 1955).
[3] de Broglie, L., *La Physique Quantique, Restera-t-elle Indéterministe?*, Gauthier-Villars, Paris, 1953 (Eng. tr. in L. de Broglie, *New Perspectives in Physics*, Basic Books, N.Y. 1962, p. 83).
[4] Bohm, D. and Bub, J., *Rev. Mod. Phys.* **38** (1966), 453.
[5] Wiener, N. and Siegel, A., *Phys. Rev.* **91** (1953), 1551; *Nuovo Cimento, Suppl.* **2** (4) (1953), 982; A. Siegel, *Synthese* **14** (1962), 171, Chapter 5 of N. Wiener *et al.*, *Differential Space, Quantum Systems and Predictions* (ed. by B. Rankin), M.I.T. Press, Cambridge, 1966.
[6] Jauch, J. M. and Piron, C., *Helv. Phys. Acta* **36** (1963), 827.
[7] Gleason, A. M., *J. Math. Mech.* **6** (1957), 885.
[8] Kochen, S. and Specker, E. P., *J. Math. Mech.* **17** (1967), 59.
[9] For a thorough review of both aspects, see F. J. Belinfante, *A Survey of Hidden-Variable Theories*, Pergamon, 1973.

[10] Bell, J. S., *Rev. Mod. Phys.* **38** (1966), 447.
[11] Papaliolios, C., *Phys. Rev. Letters* **18** (1967), 622.
[12] Belinfante, F. J., *op. cit.*, p. 245 ff.
[13] Einstein, A., Podolsky, B., and Rosen, N., *Phys. Rev.* **47** (1935), 777.
[14] Bell, J. S., *Physics* (Long Island City, N.Y.) **1** (1965), 195.
[15] Bohm, D. and Aharonov, Y., *Phys. Rev.* **108** (1957), 1070.
[16] Bohr, N., *Phys. Rev.* **48** (1935), 696.
[17] Einstein apparently was prepared to do this. See A. Shimony, 'Experimental Test of Local Hidden-Variable Theories', in *Foundations of Quantum Mechanics*, Il Corso (ed. by B. d'Espagnat), Academic Press, N.Y., 1971, p. 182ff.
[18] Clauser, J. F., Horne, M. A., Shimony, A., and Holt, R. A., *Phys. Rev. Letters* **23** (1969), 880.
[19] Horne, M. A., 'Experimental Consequences of Local Hidden-Variable Theories' (Ph.D. Thesis), Boston University, 1970, unpublished.
[20] Wigner, E. P., *Am. J. Phys.* **38** (1970), 1005.
[21] Belinfante, F. J., *op. cit.*, p. 294 ff.
[22] Bell, J. S., in *Foundations of Quantum Mechanics* (ed. by B. d'Espagnat), Academic Press, N.Y., 1971.
[23] Freedman, S. J., Experimental Test of Local Hidden-Variable Theories' (Ph.D. Thesis), University of California, Berkeley 1972 (LBL Report 391), unpublished.
[24] Freedman, S. J. and Clauser, J. F., *Phys. Rev. Letters* **28** (1972), 938.
[25] Holt, R. A., 'Atomic Cascade Experiments' (Ph.D. Thesis), Harvard University, 1973, unpublished.
[26] Holt, R. A. and Pipkin, F. M., to be published.
[27] Hanna, R. C., *Nature* **162** (1948), 332.
[28] Bleuler, E. and Bradt, H. L., *Phys. Rev.* **73** (1948), 1398.
[29] Vlasow, N. A. and Dzehelepov, B. S., *Dokl. Akad., Nauk SSSR* **69** (1949), 777.
[30] Wu, C. S. and Shaknov, J., *Phys. Rev.* **77** (1950), 136.
[31] Hereford, F. L., *Phys. Rev.* **81** (1951), 482.
[32] Bertolini, G., Bettoni, M., and Lazzarini, E., *Nuovo Cimento* **2** (1955), 661.
[33] Langhoff, H., *Z. Physik* **160** (1960), 186.
[34] Kasday, L. R., in *Foundations of Quantum Mechanics* (ed. by B. d'Espagnat), Academic Press, N.Y., 1971.
[35] Faraci, G., Gutowski, D., Notarigo, S., and Pennisi, A. R., preprint, Istituti di Fisica Dell'Universita Di Catania, PP/394.

JEAN GRÉA

PRE-QUANTUM MECHANICS. INTRODUCTION TO MODELS WITH HIDDEN VARIABLES

ABSTRACT. Within the context of formalisms of hidden variable type, we consider the models used to describe mechanical systems before the introduction of the quantum model. We give an account of the characteristics of the theoretical models and their relationships with experimental methodology. We then study in succession the models of analytical, pre-ergodic, ergodic, stochastic, statistical and thermodynamic mechanics. At each stage, the physical hypothesis is enunciated by postulate corresponding to the type of description of the reality of the model. Starting from this postulate, the physical propositions which are meaningful for the model under consideration are defined and their logical structure is indicated. It is then found that on passing from one level of description to another, we can obtain successively Boolean lattices embedded in lattices of continuous geometric type, which are themselves embedded in Boolean lattices. It is therefore possible to envisage a more detailed description than that given by the quantum lattice, and to construct it by analogy.

1. INTRODUCTION

The establishment of a model is a fundamental step taken by the physicist in the comprehension of reality. The methodological axiom sustaining this step is that any model is a stage in technique and its only purpose is to be dismantled [5]. Since its origin and up to its axiomatic formulation, the quantum model has shown itself in one feature as irreducible to the previous models (classical, statistical, thermodynamic, etc....) and incompatible with any extension of a 'complete' model (in the Einstein sense).

The aim of this work is to examine the series of pre-quantum models in order to indicate the conditions under which they describe physical systems and the mathematical expression of these conditions. On this occasion we will see that the principle of existence of 'hidden variables' is not recent, and that its formulation is exhibited by a limited number of structures. Starting from these we can envisage different possibilities for the 'hidden variables' of a quantum model.

In our examination of each of these models, we will attempt to give a precise formal expression of the set of propositions linked with it and to define the isomorphism linking it with the set of experimental propositions which corresponds to it.

The order of increasing restrictions imposed on the experimental

M. Flato et al. (eds.), Quantum Mechanics, Determinism, Causality, and Particles, 61–103. *All Rights Reserved. Copyright © 1976 by D. Reidel Publishing Company, Dordrecht-Holland.*

conditions will lead us to consider in turn the analytical model, the ergodic, stochastic statistic and then the thermodynamic models.

2. Rational Non-Relativistic Mechanics as a Model

A. *Definition*

This model consists of the isolation of a part D of reality with regard to certain *a priori* forms of our perception. By universal agreement, we allow to each concrete realization of these forms, numbers which will be the coordinates $x_\nu(\nu=1, N+1)$ of the realization. The whole set of possible coordinates generates a manifold M, starting from which the formal study of the system and of its evolution are assumed possible [1, 28] and unambiguous within D.

First Postulate of the Model: Stability for Direct Composition

If $x_\nu(\nu=1, N_1)$ and $y_\mu(\mu=1, N)$ are the coordinated associated respectively with the system S_1 and S_2 concerned with the model, there exists a set of co-ordinates (x_ν, y_μ) associated with a system denoted by $S_1 \cup S_2$ for which the model is valid. Conversely, if S is associated with $\{z_k\}$ and $\{x_\nu\}$ is a sub-set of $\{z_k\}$ associated with S_1 by the model, there exists S_2 with $S=S_1 \cup S_2$ and $\{y_\mu\}$ with $\{z_k\}=\{x_\nu\}\otimes\{y_\mu\}$.

Second Postulate: the Equation of Evolution

Within D the evolution of the system is associated with a group of M automorphisms with one parameter: U_t. Locally on M, with the chart $\{\mathcal{O}, x_\nu\}$, this evolution is then written by n-upple

$$x_\nu \in \mathcal{O} \subset M, \qquad \nu=1, n; \qquad \frac{\mathrm{d}X_\nu}{\mathrm{d}t}=X_\nu(x, t, \alpha), \tag{1}$$

where the X_ν are real uniformly continuous functions in the open region $\mathcal{O}$ and with respect to α where α represents a set of parameters which are essential and associated with the physical constants of the system. The X_ν are the Lie generators of U_t on $\mathcal{O}$.

Third Postulate: T-Invariance

Since the system is closed, the description is invariant by reversal for the motion. There exists an involution T on M where:

$$TTP = P,$$

where P is any point whatsoever of M, where

$$\begin{aligned} &x(P) = x_v \\ &TU_t = U_{-t}T. \end{aligned} \tag{2}$$

There are two consequences. The model defines an order of refinement of the description by the first postulate. A description is more refined if it connects a manifold with a larger range of dimension, that is if the number of 'constituents' and hence interactions for a given systems, increases.

The notion of the state of the system is a derived notions. Since the fundamental data is the *a priori* forms and the physical constants, known as the fundamental magnitudes of the system, the state is associated with a set of values of these magnitudes.

The fact of isolating the system within D is denoted by $\{X_v\}$ which is usually called the model of the system.

B. *Formalisms*

(a) *Newton's formalism.* Systems are made up of points. The X_v do not explicitly depend on the time.

Resolution of (1) leads to the solution (6, d)

$$x = \chi(\mathring{x}, \mathring{t}, t), \qquad \text{where} \quad \mathring{x} = \chi(\mathring{x}, \mathring{t}, \mathring{t})$$

Reversibility is written simply as $\mathring{x} = \chi(x, t, \mathring{t})$. The description is then given by the N first integrals $\mathring{x}(s, t, \mathring{t})$.

Two systems will be equivalent for (1) if there exists a diffeomorphism between their manifold which associate the system of generators of the first with the second.

(b) *Euler's formalism.* The existence of links leads to systems of generalized co-ordinates and the well-known equations:

$$\frac{\mathrm{d}}{\mathrm{d}t}\left(\frac{\partial T}{\partial \dot{q}_i}\right) - \frac{\partial T}{\partial q_i} = Q_i(q_i \mathring{q}_i t) \quad \text{with} \quad T = \tfrac{1}{2} \sum g_{ij} \dot{q}_i \dot{q}_j \quad (i, j = 1, n). \tag{3}$$

Since the co-ordinates are no longer directly defined by observation it is essential to characterise the equivalence of two descriptions, that is the identification for this model of two systems.

It is shown that if:

$$\frac{d}{dt}\left(\frac{\partial \mathcal{T}}{\partial \dot{\eta}_i}\right) - \frac{\partial \mathcal{T}}{\partial \eta_i} = D_i(\eta_i \dot{\eta}_i t) \quad \text{where} \quad \mathcal{T} = \tfrac{1}{2} \sum \gamma_{ij} \dot{\eta}_i \dot{\eta}_j \tag{4}$$

describes the same system as (3), then:

$$D_i = cQ_i \quad \text{and} \quad \gamma_{ik} = cg_{ik}$$

and there exists a bijective transform $\eta = \varphi(q)$ enabling us to pass from one description to the other.

The change $\mathcal{T} \to \mathcal{T} \circ \varphi \equiv \mathcal{T}^*$ enables us to express (4) in the q frame and by identification:

$$\begin{Bmatrix} j & & k \\ & i & \end{Bmatrix}_\gamma = \begin{Bmatrix} j & & k \\ & i & \end{Bmatrix}_g \tag{5}$$

which are differential equations of first order in $g_{\mu\nu}$ and $\gamma_{\mu\nu}$ where

$$g_{\beta\alpha}[q(o)] = \gamma_{\beta\alpha}[q(o)].$$

The converse is clear

The model therefore associates in a unique manner with the physical system, the $n+1$ dimensional manifold, a chart of which is $(q_i, \dot{q}_i, t)$ up to the φ equivalence.

(c) *Lagrange-Hamilton formalism.* We restrict ourselves to the class of interactions which make a perfect differential dW for fixed t of the Pfaff form:

$$\pi_d = \sum_{\nu=1}^{n} X_\nu(g, t)\, dq_\nu = dW. \tag{6}$$

The interactions are derived from a potential. This fact introduces a local property of the system, that of kinetic reversibility. The model gives a local description of quasi-static system.

The equations of motion are obtained in a well-known manner

$$\frac{\delta L}{\delta q_i} = 0, \qquad L = T + W + \frac{df}{dt}(q) \quad \text{Euler's equations} \tag{7}$$

and in the change of variables: $\dot{q}_i = \dot{q}_i(p)$, $p_i(t) = \partial L/\partial \dot{q}_i$ where $i = 1, n$ for a non-zero Hessian of L:

$$\Delta = |\partial^2 L/\partial q_i \partial q_j| \neq 0$$

$$\mathbf{V} = (\nabla_{p_i} H, -\nabla_{q_i} H) \quad \text{Hamilton's equations} \tag{8}$$

which are the Pfaff equations associated with:

$$\omega_d = \sum_i^n p_i \, \mathrm{d}q_i - H \, \mathrm{d}t. \tag{9}$$

With the same aim as before, we must give an equivalence relationship concerning the different descriptions of the same system or concerning the identification of two systems by the model.

It is well known that $(q, p, H, t) \sim (Q, P, H, T)$ if there exists a function $G(q, P, H, t)$ where:

$$\delta\omega_d - \mathrm{d}\omega_\delta = 0 \leftrightarrow \sum p_i \, \mathrm{d}q_i - H \, \mathrm{d}t - \sum P_i \, \mathrm{d}Q_i + K \, \mathrm{d}t = \mathrm{d}G. \tag{10}$$

These transforms do not apply to functions explicitly dependent on time. If we wish to conserve the separation $(p, q) \otimes t$ in the equivalence, we limit ourselves either to contact transform with Pfaff equations associated with $\omega'_d = \sum p_i \, \mathrm{d}q_i$, or with infinitesimal transforms of the first type. In all cases, these transforms conserve Poisson's relationship:

$$(\ ,\ ) = \sum_{i=1}^n \left(\frac{\partial}{\partial q_i}, \frac{\partial}{\partial p_i} - \frac{\partial}{\partial p_i}, \frac{\partial}{\partial q_i} \right).$$

A practical form of the equivalence transforms for the Hamilton model is that which is given by

$$\mathfrak{J}^t S \mathfrak{J} = sS$$

$$\mathfrak{J} = \frac{\partial Z}{\partial z} \quad \text{where} \quad S = \begin{pmatrix} 0 & -\mathbb{I}^n \\ \mathbb{I}^n & 0 \end{pmatrix} \begin{matrix} z \equiv (q, p) \\ Z \equiv (Q, P) \end{matrix} \tag{11}$$

The $\mathfrak{J}$ then provide a representation with $2n$ dimensions of the sympletic group, on $\Gamma = (p, q)$.

In conclusion, all the models considered associate with the physical system a compact part M of a manifold of a space Γ known as 'phase spaces'. The state is defined by a point of $M \times \mathbb{R}_t$; the evolution of the system is provided by the group of transforms defined in (1), (3), (4), (7) and (8) which belong by definition to the equivalence transforms.

3. Experimental propositions

In the exchange of information between the theorist and the experimenter, the latter makes use of a certain number of propositions $\mathscr{L}_{\mathscr{E}}$ relative to a set $\mathscr{E}$ of measuring equipment and $\mathscr{S}$ of physical systems considered. The propositions have truth values which are:

– independent of the models,

– independent of time although $\mathscr{E}$ and $\mathscr{S}$ do depend on time (time is here taken to mean epoch).

Finally the collection of propositions $\mathscr{L}_{\mathscr{E}}$ forms a sub-group of a set $\mathscr{L}$ which is finite at most.

So as to examine more closely the structure of these proposition [28], let us characteristise a measuring device.

(a) *The Measuring Equipment* $\mathscr{A} \in \mathscr{E}$

This is by definition associated with an observable quantity A of at least $\mathscr{S}' \simeq \mathscr{S}$ The possible values belong to a compact set $\mathbb{R}_{\mathscr{A}} \subset \mathbb{R}$ of real numbers. The sensitivity is associated with $\Delta a \in \mathscr{B}(\mathbb{R})$, the Borel of $\mathbb{R}$. It is the absolute value of Δa written $|\Delta a|$. Then: for a given $\mathscr{A}$, $\exists \{a_i\}$ where $i = 1, 2, 3, \ldots, n$, $a_i \in \mathbb{R}$ and:

$$\bigcup_{i}^{n} \Delta a_i = \mathbb{R}_{\mathscr{A}} \quad \Delta a_i = [a_i - \Delta a, a_i + \Delta a] \quad \Delta a_i \cap \Delta a_j = \emptyset; \quad \forall i \neq j$$

that is: there is a scale of measurement associated with each equipment.

If $E = \bigcup_{i=1}^{n} \Delta a_i$, the experimental proposition (the value of the observable quantity A lies within E of the scale) is written $Q_E^{\mathscr{A}}$. More precisely we have a class of equivalence of measuring equipments written $\mathscr{A}$. Two equipments are equivalent if there exists an isomorphism over the range of their questions. For a single magnitude, two equipments of different precisions are not equivalent. The order relationship:

$$Q_E^{\mathscr{A}} \geqslant Q_F^{\mathscr{A}} \Leftrightarrow F \subset E \tag{1}$$

the significance of which is evident, impresses on the set $Q_E^{\mathscr{A}}$ a structure of upper-half lattice [39a] with bound $\mathbb{I} = Q_{\mathbb{R}_{\mathscr{A}}}^{\mathscr{A}}$ and a lower bound $Q_{\emptyset}^{\mathscr{A}} = 0$. The involution:

$$Q_E^{\mathscr{A}} \to \tilde{Q}_E^{\mathscr{A}} = Q_{\mathbb{R}_{\mathscr{A}} - E}^{\mathscr{A}} \tag{2}$$

is a semi-complementation. The atoms are the $Q^{\mathscr{A}}_{\Delta a_i}$. We pass to the lattice structure by $A^{\mathscr{A}}_E \cap Q^{\mathscr{A}}_F = (\tilde{Q}^{\mathscr{A}}_E \cup \tilde{Q}^{\mathscr{A}}_F)$. The set of $\tilde{Q}^{\mathscr{A}}_F$ is then a complete atomic Boolean lattice.

The experimental observable A is then a σ-homorphism of $\mathscr{B}(\mathbb{R}_{\mathscr{A}})$ on $\mathscr{L}_{\mathscr{A}}$.

Note. For one and the same observable quantity A let us consider the finite set of equipments $\mathscr{A}_i(\mathbb{R}_A, \Delta^i a)$ of differing degrees of precision. The set of $Q^i_{E_{n,i}}$ is then a lattice, $\mathscr{L}_A$ with the same bounds as those of $\mathscr{L}_{A_i}$ and, $\forall i$, $\mathscr{L}_{A_i}$ is a sub-lattice of $\mathscr{L}_A$. $\mathscr{L}_A$ is atomic-orthocomplemented, non modular, and of finite length. The notion of an observable quantity is linked with the equivalence class on the $\mathscr{L}_{\mathscr{A}_i}$ by

$$\mathscr{A} \sim \mathscr{B} \leftrightarrow \exists \mathscr{L}_C; \qquad \mathscr{L}_{\mathscr{A}} \underset{T}{\subset} \mathscr{L}_C \quad \text{and} \quad \mathscr{L}_{\mathscr{B}} \underset{T}{\subset} \mathscr{L}_C \tag{3}$$

$\mathscr{L}_A$ is usually considered as a Boolean lattice of the observable events. We can in particular obtain this result by again taking the related lattice to have the greatest sensitivity, or by limiting ourselves to one single sub-lattice $\mathscr{L}_{\mathscr{A}_i}{}^*$ (p. 101).

(b) $\mathscr{S} = \{\Sigma\}$. *The Systems*

Σ corresponds experimentally to an object isolated by a set of experimental arrangements, each process being necessarily reproducible. Σ is a class of equivalence of such processes. The elements of the class $\sigma \in \Sigma$ are realizations of the system. For each Σ there exists a sub-set $F_\Sigma \subset \mathscr{E}$ such that for all $\mathscr{A}_i \in F_\Sigma$ we have:

$$\begin{aligned} &(\sigma, Q^{\mathscr{A}_i}_E) \to \{0, 1\} \quad \forall E \subset \mathbb{R}_A \\ &F_\Sigma = \{\mathscr{A}_i, \forall \sigma \in \Sigma, \quad (\sigma Q^{\mathscr{A}_i}_{\mathbb{R}_{\mathscr{A}_i}}) = 1\}. \end{aligned} \tag{4}$$

The equivalence class F_Σ with respect to (3) is known as the set of experimentally observable quantities of Σ, written $\mathscr{O}_\Sigma$.

We define the lattice $\mathscr{L}_\Sigma$ of experimental propositions relative to Σ as the free product [37b] of the $\mathscr{L}_{A_i}$ for $A_i \in \mathscr{O}_\Sigma$. $\mathscr{L}_\Sigma$ is generated by union and intersection of all the n-uples $[Q^{A_i}_{E_{m_i}}]$ $i = 1, N$, $A_i \in \mathscr{O}_\Sigma$, $E_{m_i} \neq \emptyset$. $\mathscr{L}_\Sigma$ is boolean σ complete and atomic. The atoms are propositions of type:

$$[\Delta a_{m_1} \dots \Delta a_{m_n}] \quad \text{where} \quad |\Delta a_{m_i}| = \inf |\Delta a_i|. \tag{6}$$

For all A_i of $\mathcal{O}_\Sigma$, $\mathcal{L}_{A_i}$ is ortho-isomorphic with a sub-lattice of $\mathcal{L}_\Sigma$. The surjective homomorphism of $\mathcal{L}_\Sigma$ on $\mathcal{L}_{A_j}$ will be written as Φ_j

$$\Phi_j\colon [Q^{A_i}_{E_{m_i}}] \to Q^{A_j}_{E_{m_j}} \tag{7}$$

By construction, all the experimental propositions concerning Σ are contained in $\mathcal{L}_\Sigma$.

(c) $S_\Sigma = \{s\}$ *Set of States of* Σ

Experimentally, at each realization $\sigma \in \Sigma$, the measurements of the $A_i \in \mathcal{O}_\Sigma$ provide one single result which has no physical significance. Reproducibility therefore plays an integral part in the definition of the realizations of a system.

Out of a set of such realizations (called a 'physical experiment') we can give a coherent definition of a statistic [14] based on the frequencies of the results obtained. If the set of experimental conditions is defined in such a manner that this statistic is reproducible, we say that the system Σ has been produced experimentally in a state s which is well defined; s is thus associated with an equivalence class of $\sigma \in \Sigma$.

By the definition of a statistic we have:

$$\begin{array}{ll} 0 \leqslant s \cdot Q^{(A_i)}_{(E_i)} \leqslant 1 & \forall Q^{(A_i)}_{(E_i)} \in \mathcal{L}_\Sigma \\ s \cdot 0 = 0 & s \cdot \mathbb{I} = 1 \end{array} \tag{8}$$

if

$$Q^{(A_i)}_{(E_i)} \leqslant \tilde{Q}^{(A_j)}_{(E_j)} \quad \forall i \neq j \qquad s \cdot [\bigvee_i Q^{(A_i)}_{(E_i)}] = \sum_i [s \cdot Q^{(A_i)}_{(E_i)}]$$

s is an isotonic measure of probability on $\mathcal{L}_\Sigma$:

$$s \in S_\Sigma\colon Q^{(A_i)}_{(E_i)} \leqslant Q^{(A_j)}_{(E_j)} \mapsto s \cdot Q^{(A_i)}_{(E_i)} \leqslant s \cdot Q^{(A_j)}_{(E_j)} \tag{9}$$

and $s \cdot \Phi_j$ is an isotonic measure of probability on $\mathcal{L}_{A_j}$ $\forall j$ with Φ_j defined in (7).

(d) *The Physical Propositions of* Σ

The elements are those of $\mathcal{L}_\Sigma$ on which new order relations have been established by experiment, compatible with those of $\mathcal{L}_\Sigma$ and called 'experimental laws'. The elements concerned are called [13] 'elements of reality'. If a law is denoted by p, then:

$$p = [\{(Q^{A_i}_{E_{i_j}}, Q^{A_k}_{E_{k_n}})\}\ \exists s \in S_\Sigma \quad \text{with} \quad s \cdot Q^{A_i}_{E_{i_j}} = 1 \to s \cdot Q^{A_k}_{E_{k_n}} = 1]. \tag{10}$$

For all realizations σ of S we have the above implication and the $Q^{A_k}_{E_{k_n}}$ are elements of realities.

DEFINITION. *We say that* $Q^{A_i}_{E_{i_j}}$ *implies* $Q^{A_k}_{E_{k_n}}$ *weakly if there exists* $s \in S_\Sigma$ *for which the correlation coefficient* $\rho(Q^{A_i}_{E_{i_j}}, Q^{A_k}_{E_{k_n}}) = 1$ *for the statistic defined by s. Clearly a strong implication implies a weak implication.*

The reduction of $\mathscr{L}_\Sigma$ by the laws p is the lattice of physical propositions of Σ or the experimental model of Σ written L_Σ.

4. Complete Total Physical Theory

The structure of L_Σ is in general very complex [23]. What we wish to show is that the aim of the physicist is to obtain a Boolean lattice, or to obtain a theory with maximum predictive power. For this purpose we express within L_Σ two characteristics defining all experimental projects:

1. If $a<b$, $\quad \exists c \in L_\Sigma \quad a' \wedge b = c$ and $c \vee a = b$.

2. $\forall a \in L_\Sigma, \quad a = \bigvee_{\omega_a} \alpha_i; \quad \forall i, \alpha_i \succ 0$ where $\omega_a = \{\alpha_i \leqslant a\}$.

1. Means that any phenomenon which is a consequence of a, but is reproducible independently of a is a consequence of a factor c of L_Σ independent of a, maximum in the sense that we have $sb=1 \rightarrow sa=1$ *or* $sc=1$, we have the Boolean lattice.

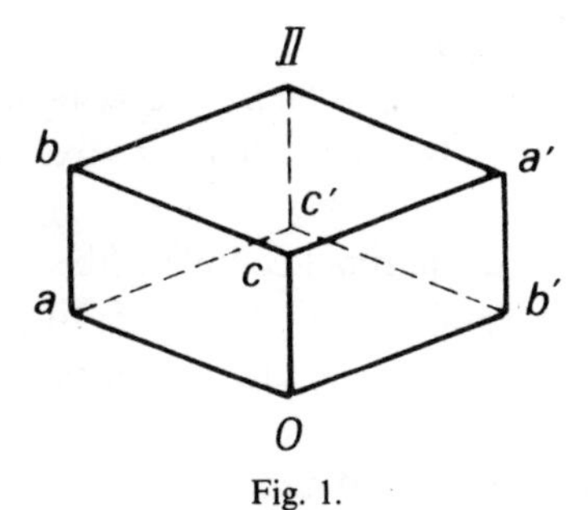

Fig. 1.

An equivalent formulation [35] is a $a<b$, $a' \wedge b = 0 \rightarrow a = b$.

Two experimental propositions are identical if one implies the other and the other never takes place without the first one taking place (we take the quotient of L_Σ by the equivalence relationship:

$$a<b \quad \text{and} \quad a' \wedge b=0 \qquad (b'<a') \qquad b\sim a.$$

2. Means that within the framework $\mathcal{O}_\Sigma$ (that is a given time) the description is complete, or every proposition is implied by a defined sub-set of elementary propositions (atomic).

We will further suppose that there exists $P_\Sigma \subset S_\Sigma$ where:

$$\forall s' \in S_\Sigma, \exists \{\lambda_i \in \mathbb{R}\} \quad \text{and} \quad \exists \{s_i \in P_\Sigma\};$$
$$\forall a \in \mathcal{O}_\Sigma, \sum_i \lambda_i s_i(a) = s'(a) \sum_i \lambda_i.$$

Two cases now arise:

(a) Whatever $s \in P_\Sigma$ and $a \in L_\Sigma$ may be, the experimentor can predict $sa=0$ or $sa=1$. This means that he has been able experimentally to associate an atom of L with every s which he can prepare (that is a set of values of all $C \in \mathcal{O}_\Sigma$ with maximum precision) and that he has defined ω_a, $\forall a \in L_\Sigma$. The lattice gives a complete description, or alternatively we can say that the system is completely determined. This has frequently occurred in history. It is then trivial to show that L_Σ is Boolean and P_Σ a base of S_Σ (pure states).

(b) Whatever $s \in S$ and $a \in L_Z$ may be, $s \cdot a \in [0, 1]$ can be predicted. Two eventualities exist:

- either the sub-lattice $\lambda_s \subset_T L_Z$ where $\lambda_s = \{a \in L_Z;\ s \cdot a = 1 \text{ or } sa=0\}$ is above an atom of L_Σ and $\forall \alpha \in L_Z$, $\alpha \succ 0$, he is able to produce s_α such that λ_{s_α} is generated by α, that is he has a catalogue of 'pure states' and all $s \in S_\Sigma$ where $\lambda_s = \emptyset$ will be interpreted as a mixture of pure states (this is the quantum case),
- or $\forall s \in L_Z$, $\lambda_s = \emptyset$, the system is not complete (this is the stochastic case).

In the two cases the description is clearly not total, and the aim of the experimentor has not been reached, since he has not got complete control of the system.

There are three solutions left to him:

- to increase the precision of his measuring instruments and to complicate his experimental arrangements so as to increase L_Σ and S_Σ (new filters [23]),
- to assume that $\mathcal{O}_\Sigma$ is not sufficient and to ask the theoretician to in-

vestigate the results concerning the $s \in S_\Sigma$, $s \notin \lambda_s$ to express them in functions of new observable quantities (hidden variables),

– to assume that Σ itself has not been well defined, and that in fact Σ is a complex (internal structures).

Conclusion

The aim of physics in every case is to obtain description (a), that is a Boolean lattice. It is certain that such a lattice is not associated with one single model, but if L_Σ must be Boolean, the lattice $\tilde{L}_\Sigma$ of the propositions of a physical model must be regarded as embedded in L_Σ the embedding being not necessarily given by a morphism.

Conversely, the conditions enabling us to state that the limit of a series of lattice of propositions concerning successive models of a physical system really does exist, and further that it is Boolean, are far from having been stated. We can indicate one case where this converse is verified.

Let us consider a family of complete lattice, $\tilde{L}^i_\Sigma$ orthocomplementary and atomic. For each $\tilde{L}^i_\Sigma$ we have a set of states $\tilde{S}^i_\Sigma$ complete in the sense of:

$$\{(a, b) \in \tilde{L}^i_\Sigma\}, \qquad \{sa \leqslant sb \ \forall s \in S^i_\Sigma\} \rightarrow a \leqslant b. \tag{1}$$

The states separate the propositions concerning Σ which is coherent with Section 3(a). Further, for each i there exists a lattice of experimental propositions L^i_Σ Boolean (it is sufficient to take $\mathscr{L}_\Sigma$ as defined in Section 3(a)) and a set of states S^i_Σ experimental and complete; to sum up, we have a logic $(\tilde{L}^i_\Sigma, \tilde{S}^i_\Sigma)$ attached to the system Σ by the model i and correctly describing the logic (L^i_Σ, S^i_Σ) of the experiments concerning Σ.

HYPOTHESES. *For all i there exists* $(\mathscr{L}^i, \mathscr{S}^i)$ *where* $L^i_\Sigma \subset_T \mathscr{L}^i$ *(respectively* $\tilde{L}^i_\Sigma \subset_T \mathscr{L}^i$*) and:*

$$\forall \sigma \in \mathscr{S}^i \exists s \in S^i_\Sigma \quad (\text{resp. } \tilde{s} \in \tilde{S}^i_\Sigma) \tag{2}$$

with:

$$\forall a \in L^i_\Sigma \quad (\text{resp. } \tilde{a} \in \tilde{L}^i_\Sigma)$$

$$\sigma a = sa \quad (\text{resp. } \sigma \tilde{a} = \tilde{s} \tilde{a}).$$

This hypothesis shows by the structures defined in Section 3, the first postulate of the mechanical model defined in Section 2: the partition of reality into a system Σ and the remainder is purely artificial.

Further, to say that $(\tilde{L}_\Sigma, \tilde{S}^i_\Sigma)$ is the model of (L^i_Σ, S^i_Σ) means that there exists a surjection φ of a non-empty part B of L_Σ on $\tilde{L}^i_\Sigma$ with:

$$\sigma a = \sigma\varphi(a) = \sigma[\varphi(a) \wedge a] = \tilde{s}\varphi(a). \tag{3}$$

THEOREM. *If the description is maximal, the logic corresponding to it is Boolean.*

If the description is maximal, then $(\mathscr{L}_\Sigma, \mathscr{S}_\Sigma) = \mathrm{Sup}_i(\mathscr{L}^i_\Sigma, \mathscr{S}^i_\Sigma)$ for the order inclusion relation. Then there exists $(L_\Sigma S_\Sigma) \subset (\mathscr{L}_\Sigma, \mathscr{S}_\Sigma)$ experimental logic associated with $(\mathscr{L}_\Sigma, \mathscr{S}_\Sigma)$ which for (2) and (3) may be considered as the logic of the complete model of Σ.

There exists therefore φ as a surjection of a part $L' \subset L$ on $\mathscr{L}$ and

$$\forall \sigma \in \mathscr{S}_\Sigma \qquad \forall a \in L' \qquad \sigma\varphi(a) = \tilde{s}\tilde{a} \Rightarrow \sigma = \tilde{s}. \tag{4}$$

Now $\mathscr{S}_\Sigma$ is complete, therefore $\varphi(a) = a \rightarrow \varphi = \mathbb{I} \qquad L' = L = \mathscr{L}$.

5. Classical logic

If we reconsider the results of Section 2, we find [6, a] that to a system Σ, classical mechanics associated a complete atomic Boolean $(\tilde{L}_\Sigma \tilde{S}_\Sigma)$. Whatever the formalism, the equivalence classes are defined by the dimension of the manifold M the points of which are the atoms of $\tilde{L}_\Sigma$. The observable quantities are the elements of a commutative Lie algebra of infinite dimension, the representations of which within the various formalisms of Section 2, are Borel functions on $\tilde{L}_\Sigma$ with range in $\mathbb{R}$. It is known [31, 32] that a realization in terms of a Hilbert space $\mathscr{H}$ of squared functions summable on M (provided with the U_t invariant measure (Section 2, Equation (2)) where the observable are operators, is then commutative and $\mathscr{H}$ is simply the direct integral of Hilbert space of one dimension. (This is a formalism known as continuous super-selection [35a].) This model corresponds to a determinist experimental logic $(L_\Sigma S_\Sigma)$. For this model to be total, it is necessary and sufficient for the definite isomorphism φ (Section 4, Equation (2)) to exist for $\mathcal{O}_\Sigma$ in the limit $\Delta a = 0$ where $\Delta a = \inf_{\mathscr{A}_i} \Delta^i a, \forall \mathscr{A}_i \in A \in \mathcal{O}_\Sigma$ (cf. Section 3(a)). In other words its is necessary and sufficient that:

(a) The proposition associated with the points of a sub-set which is countable and dense within M can be experimentally obtained (infinite precision)

(b) for this experimental lattice φ exists.

Proposition (a) has been demonstrated [32b] under the condition that the interval of precision Δa is associated with a sphere $K(P, \varepsilon)$ of centre P, radius ε and $A(P') \in \Delta a$, $\forall P' \in K$ with the Lebesgue measure on M and that the measure of M is finite.

With regard to (b) it is well known experimentally that beyond a given degree of precision, φ no longer exists (the lattice of the model of microscopic propositions is no longer Boolean [35b]).

Following this, numerous attempts to increase $\mathcal{O}_\Sigma$ and to immerse (L_Σ, S_Σ) in $(\tilde{L}_\Sigma, \tilde{S}_\Sigma)$ take place without success. For these attempts, the theorems of non-existence of hidden variables [35c, 30] are valid. We shall see that this type of extension is not that of the physical models which have succeeded classical mechanics. It is not surprising that the quantum model is resistant to it.

6. THE PRE-ERGODIC MODEL

To the postulates of Section 2 we add:

(W) POSTULATE *S. For all Σ the preparation of Σ cannot be instantaneously realized*

$$\forall \Sigma; \qquad \mathcal{O}_\Sigma = \{\mathcal{O}^w_\Sigma, t, \Delta t\},$$

where $\mathcal{O}^w_\Sigma$ is the set of classes of measuring equipment relative to Σ and not equivalent to a clock.

In order to construct $(\tilde{L}^w_\Sigma, \tilde{S}^w_\Sigma)$ we start from the classic lattice Section 5. Those of the atoms which are compatible with (W) are of type $Q^{(A_i),t}_{(a^i{}_n),\, \mathbb{R}^+}$ that is the minimal propositions of the A_i where $A_i = a^i_n \forall t$.

Physical quantities of these types are the independent time first integrals of Σ. We suppose that all the A_i are given by continuous functions $\varphi_i(P)$ on M which are $\tilde{\mu}$ measurable ($\tilde{\mu}$ being the U_t invariant Lebesgue measure on M).

(a) *Parametrization by the First Integrals*

Let us consider a vector $\Phi(z, t)$ of dimension $2m(m<n)$ with rank of $\text{grad}_z \Phi = 2m$. Then (I) $\Phi = \Phi(z, t) = 0^{2m}$ defines, t locally, a manifold Γ of dimension $2(n-m)$. We can always choose a $2(n-m)$ vector ψ of compo-

nent ψ_i and a $2n$ vector F such that $z = F(\psi, t)$ is local parameter of Γ and that $J = \text{grad}_\psi F$ is of rank$2(n-m)$.

If all the solutions $z(t)$ of Hamilton's equations:

$$S\dot{Z} = \text{grad}_z H$$

which are in Γ for one t, remain for all t, the relationships (1) are known as invariant and the solutions are given by:

$$SJL = -S\partial_t F + \text{grad}_z H$$
$$L(\psi, t) = \dot{\psi}$$

with:

$$S = \begin{pmatrix} 0 & -\mathbb{I} \\ \mathbb{I} & 0 \end{pmatrix}$$

(b) *Case of the First Isolating Integrals* [33]

DEFINITION. *The functions* $\varphi_i, \ldots, \varphi_{2m}: \Gamma \to \mathbb{R}$ *constitute an isolating set if*:

$\forall i,\ \varphi_i(\Gamma) = I_{c_i} \subset \mathbb{R}$	*is a connected open set*
$\forall c_i \in I_i,\ \varphi_i^{-1}(c_i)$	*is closed sub-manifold of* Γ, *with* 2n-1 *dimensions*
$\bigcap_{i=1}^{2m} \varphi_i^{-1}(c_i)$	*is closed,* $2(n-m)$ *dimensional of* Γ.

If in addition the φ_i are first integrals of (2) they are known as isolating and M_c will denote the Γ sub-manifold defined by the $\varphi_i^{-1}(c)$'s.

THEOREM (J. Dieudonné). *If there exists a map* $z = (q_1, \ldots, q_n, p_1, \ldots, p_n)$ *of* Γ *in the neighbourhood* $O(P) \subset \Gamma$, *such that:*

$$J = \left|\frac{\partial \varphi_i}{\partial z_k}\right|_{z(P)} \quad \text{is of maximum rank} \tag{3}$$

the φ_i *form an isolating set:*

– *the application* $\Gamma \to \mathbb{R}^{2m}$ *is a submersion* χ,

– *the differential system induced on* M_c *by* χ *starting with Hamilton's equations is Hamiltonian.*

We consider the maximal set of such independent φ_i, $2m$ in number.

Then:

$$\forall\psi(P)\quad (U_t\psi)(P)=\psi(P)\qquad \psi(P)=f(\varphi_1,\ldots,\varphi_m),$$

where (4)

$$\forall i[\varphi_i, H](P)=0\qquad \forall P\in\Gamma.$$

Observables of this type form a Lie sub-algebra of infinite dimension, of base φ_i.

DEFINITION. *If $m<n$ the system will be known as pre-ergodic. This comes from the fact $M_c[\varphi_i=c_i]$ is not only U_t invariant but very often* [18c] *almost always indecomposable into part of non-zero measure, U_t invariant.*

The variety M_c is connected, compact and of finite measure (the hypothesis $H(z)<E_0$ and $\{z\in\Gamma H(z)=E\}$ closed in Γ is generally understood in physics and justifies the existence of a least one φ).

THEOREM [4]. *If J is of rank 2m, that is if $\{\text{grad}\,\varphi_i\}_{i=1,2m}$ is a free family, M_c is a toroid with 2m dimensions, the neighbourhoods of Γ are direct product of M_c by $\mathbb{R}^{2m}$ and there exists a base $\chi_j(z)$ of M_c such that $(\varphi,\chi)\rightarrow(z)$ is canonical.*

On M_c, this application induces an invariant measure U_t defined by [15]:

$$\mu=\frac{1}{\Omega(c)}\frac{s}{\left|\sum_{r=1}^{n}\frac{\partial\varphi_i}{\partial z_r}\frac{\partial\varphi_j}{\partial z_r}\right|^{1/2}},\tag{5}$$

where $|A|$ means $\det A$ and $\Omega(c)$ is the normalizing factor for the measure μ, $z_r\equiv(p_r,q_r)$, s is the natural measure on M_c provided with χ_j.

The algebra of the observables on M_c is a Lie sub-algebra of that of the observables of Γ. The equivalence of two systems is still given in Section 2 (Equation (11)).

(c) *Study of the Lattice $\tilde{L}^w_\Sigma$*

THEOREM. *The sub-lattice of $\tilde{L}^w_S$ relative to the propositions defined by the observables φ_i is a metric Boolean lattice.*

Let $\psi=\psi(z,t)=0^{2(\bar{n}-m)}$ be the invariant relationships on M_c which added to Φ determine the motion completely. We will suppose $\psi_i(z,t)$ component of ψ, $\tilde{\mu}$ measurable; they cannot be 'better' since Φ is maximal. These will be the first integrals of discontinuous type or infinitely multiform. Nevertheless, by using a result of C. L. Siegel [38] and of Kolmogarov [4] we know that in the space of the Hamiltonians relative to Σ, for all neighbourhoods, we have systems which are completely integrals ($n=m$) and non-integrable ($m=1$). For these latter the ψ_i are measurable non-analytic integrals. Then for Γ considered as the direct product of $M_c \times \mathbb{R}^{2m}$, ψ_i, $\tilde{\mu}$ measurable is μ measurable as a function defined on $M_c \times T$.

$$\psi_i(z, t) = \psi_i(\chi_j, c_i, t). \tag{6}$$

We make the following hypothesis:

A preparation of Σ will always be able to concern the φ_i. But since the ψ_i depend in general on time, the experimental arrangement will only be able to concern one ψ_i and will give it the value a_i at t_0. All the other ψ_j $j \neq i$ will take up their values in an interval Δa_j compatible with Δt of (W). The resulting state will provide the results:

$$\begin{aligned} &s_i Q_{a_i}^{A_i} = 1 \qquad a_i \in \mathbb{R}_{A_i} \\ &s_i Q_{a_j}^{A_j} \in [0, 1], \end{aligned} \tag{7}$$

where A_i is the observable quantity associated with ψ_i.

Since the ψ_j are functionally independent, we have:

$$\forall^j \neq i \ {}^0X_j^i = \{x \in \mathbb{R}_{A_i} : \psi_i^{-1}(x) = a_i\} \subset \{\text{open of } \mathbb{R}_{A_j}\} \tag{8}$$

and $D_i^0 = \psi_i^{-1}(a_i, t_0)$ open of M_c, $a_i \in \mathscr{B}(\mathbb{R}_{A_i})$ of precision Δa_i.

Without any further possible information we associate with s_i a σ-measure ρ_i^0 on the Borel family generated by the parts of D_i^0 compatible with (7). D_i^0 is ρ_i measurable by construction and the class of sub-sets ρ_i measurable of D_i^0 is separable. Finally $\bigcup_i D_i^0 = M_c$.

The interval Δt of (W) defines on $\mathbb{R}_{A_j}$ a family of open set $\mathscr{B}^w({}^0X_j^i) \subset \subset \mathscr{B}^w(\mathbb{R}_{A_j})$. That is

$$\delta_j^i(a_j) = \left\{ \Delta a_j' \in \mathscr{B}(\mathbb{R}_{A_j}); \psi_j \left[\bigcup_{t=t_0}^{t_0+\Delta t} U_t \psi_j^{+1}(a_j) \cap D_i^0 \right] \in \Delta a_j' \right\}$$

$$\delta_j^i(a_j) \in \mathscr{B}^w(\mathbb{R}_{A_j})$$

then for all s_i we have measure m_i on $\mathbb{R}^{2(n-m)}$ where:

$$m_i \Big[\prod_{j \neq i} \delta_j^i(a_j) \Big] = \rho_i [z, \psi_j(z) \in \delta_j^i(a_j)]. \tag{9}$$

HYPOTHESIS. $\forall t>0$, ρ_i *is uniformly continuous with respect to* μ (5)

$$\begin{aligned} &\rho_i(\mathrm{d}P_t)=\gamma_i\mu(\mathrm{d}P_t) \qquad \forall \mathrm{d}P_t\subset M_c \\ &\gamma_i\in L^1(M_c,\mu) \qquad \gamma_i>0. \end{aligned} \tag{10}$$

In the opposite case, there exists an open range ω of μ measure zero with $\rho(\omega)\neq 0$. Now $\mu(\omega)=0$, i.e. ω belongs to an open range of a sub-manifold of M_c, that is there exists at least one j where $\psi_j^{-1}(\omega)=a_j$. Then the experimental arrangement preparing $A_i=a_i$ would simultaneously prepare $A_j=a_j$. We would have locally a functional dependence of ψ_i and ψ_j. This possibility is considered to be very unlikely for globally functionally independent ψ_i (this is [21b] a common method for obtaining ergodic properties).

Consequences

The measure m_j defined by Equation (9) is a measure of probability on $\prod_j {}^0X_j^i=X^i$ associated with s_i (Section 3, Equation (8)). We will not distinguish two measurable sub-set of X^i, m_i measurable when the difference is of zero measure.

For every $i=1, 2(n-m)$, we have the decomposition $\tilde{L}_\Sigma=\tilde{L}_{\varphi_i}\otimes\tilde{L}_{\psi_j}$ where $\tilde{L}_{\varphi_k}$ is the Boolean lattice generated by the $\varphi_k=c_k$, $A_i=a_i$ (where A_i is the observable described by ψ_i) and m_i is a valuation on L_{ψ_j}, the lattice of classes of m_i measurable parts of X^i, ordered by the inclusion.

By the decomposition theorem of Γ into $M_c\times\mathbb{R}^{2m}$, $\tilde{L}_\Sigma$ is isomorphic to the lattice $L_\varphi\otimes L_\chi$. Further, $\{\psi^{-1}_{\mathbb{R}_{A_j}}\}=M_c$ since (Φ, ψ) completely determine the motion within Γ. Finally for every $a_i\in\mathscr{B}(\mathbb{R}_{A_i})$ D_i^0 is separable. Then $L_\chi\simeq L_\psi$. Finally [6c] any finite measures on a σ-algebra is a σ-additive valuation. If N is the set of elements with zero m_i measure, the lattice of classes of equivalence by N is a separable, non-atomic Boolean, metric lattice. We have therefore:

THEOREM. *For every classical controllable observable* A_i, $i=[1, 2(n-m)]$ *the lattice* $\tilde{L}_\Sigma^w$ *is decomposable into a direct product of a complete atomic Boolean lattice and a metric non-atomic Boolean lattice. The complete set of states* (7) *is the basis of the measurements on* $\tilde{L}_\Sigma^w$.

THEOREM. *The lattice of propositions of* $\tilde{L}_\Sigma^w$ *is ortho-isomorphic with a reducible modular lattice. Since the centre is an atomic lattice, the quotient is ortho-isomorphic with a sub-lattice of the lattice of the sub-spaces of Hilbert space.*

There only remains the second proposition to be proved. Let us consider the Hilbert space $\mathscr{H}(X^i, \mathscr{D}, m_i)$ of square functions m_i summable on X^i where $\mathscr{D} = \mathscr{B}(X^i)$ is the σ-algebra of the parts of X^i. Let $I(d)$ be the characteristic function for $d \in \mathscr{D}$. Let us consider the projection $f \in \mathscr{H} \rightsquigarrow I(d)\ f \in V_d$ sub-space of $\mathscr{H}$. The application $\lambda: d \rightsquigarrow V_d$ is an injection of $\tilde{L}_{\psi_j}$, $j \neq i$ in $L_{\mathscr{H}}$ lattice of the sub-space of $\mathscr{H}$.

Indeed, $\lambda(d) = \lambda(d') \leftrightarrow m_i(d \dot{+} d') = 0$, hence λ injects the equivalent m_i classes into $L_{\mathscr{H}}$. Let $\tilde{d}$ denote the d class. It will be recalled that $Q_1/Q_2 \leftrightarrow \leftrightarrow Q_1 \leqslant Q_2'$ and $\forall f, g \in \mathscr{H}$, $f \equiv g \leftrightarrow \|f - g\| = 0$. Then $\lambda(d)/\lambda(d') \leftrightarrow \tilde{d}/\tilde{d}'$ and $\lambda(d \cap d') = \lambda(d) \wedge \lambda(d')$. The application λ is an ortho-isomorphism [25].

Conclusion. Pre-Ergodic Model of 'Hidden-Variables'

Starting from a complete atomic Boolean lattice, the postulate (W) translating an experimental condition of the model has led us to a modular reducible family of lattices, each one associated with an experimental arrangement concerning a controlable observable.

The statement that all the sub-lattices of a Boolean lattice are Boolean does not prevent the passage from a Boolean lattice to an embedded modular lattice. For these extensions all the theorem forbidding the existence of hidden variables are inapplicable.

Finally we can summarize the formalism of hidden variables for $\tilde{L}_{\Sigma}^{w}$ by saying that $\tilde{L}_{\Sigma}^{w}$ is obtained from $\tilde{L}_{\Sigma}$ by introducing for every set (φ, A_i) of variables a measurement family $\rho_i(t)$ on the ergodic manifold M_c. The observables have been conserved here, but the states support a family of independent measurements $\rho_i(t)$.

The classical limit is obtained simply as $\Delta t \to 0$ then $\rho_i = \delta(\psi_i - a_i)$ and the family ψ_j being complete and simultaneously prepared $\lim_{\Delta t = 0} \tilde{L}_{\Sigma}^{w} = \tilde{L}_{\Sigma}$ atomic. It is shown indeed [9] that for all neighbourhoods of an ordinary point of Γ (that is $\tilde{\mu}$ almost everywhere) there exist $2n-1$ first integrals of the motion. Then in this neighbourhood $\rho_i = \delta(\psi_i - a_i)$ for all i.

7. The Ergodic Model

This concerns the description of the systems which exhibit a stationary character: i.e. there exists a representation where the states are defined on a compact (tore) of the variety M and where the statistics induced by these states are independent of time. For this description we retain all the preceding postulates, to which we add:

(X) POSTULATE E. *The observables of the system are the mean values averaged over time of the classical observables of the system, when these values exist*

$$A_j \in \mathcal{O}_\Sigma \to \hat{A}_j = \lim_{T\to\infty} \frac{1}{T} \int_0^T \psi_j(\mathring{z}, t)\, \mathrm{d}t, \tag{1}$$

where $\psi_j(\mathring{z}, t)$ *is the representation of* A_j *and* $(\mathring{z}, t) \in M \times t_0$.

Retaining postulate S, the states are still defined in Section 6 (Equation (7)), that is be the $\rho_i(t)$. On the other hand, the observables are no longer all those of Section 6. The observables described by the φ_i are still observables of Σ with (X) since $\varphi_i \circ U_t = \varphi_i$, $\forall t$. On the other hand, the ψ_j will give observables of Σ only on condition that:

$$\lim_{T\to\infty} \frac{1}{T} \int_0^T \psi_j(z^0, t)\, \mathrm{d}t = \hat{\psi}_j(\mathring{z}) \quad \forall \mathring{z} \in D_i^0 \tag{2}$$

exists.

LEMMA [18b]. *For all* $f \in L_1(M, \mu)$ *with a complex value:*

$$\hat{f}(x) = \lim_{T\to\infty} \frac{1}{T} \int_0^T f(U_t x)\, \mathrm{d}t \quad \textit{exists } \mu \textit{ almost everywhere } (p, p) \tag{3}$$

$\hat{f}(x)$ *is* U_t *invariant* μ $p \cdot p$, μ *summable:*

$$\hat{f}(U_t x) = \hat{f}(x) \tag{4}$$

$$\int_M \hat{f}(x)\, \mathrm{d}\mu = \int_M f(x)\, \mathrm{d}\mu \quad \text{if} \quad \mu(M) < \infty. \tag{5}$$

(a) *Case of Neighbouring Systems of Integrable Systems* ([38] and [4])

$U_t(D_i^0)$ generates an invariant sub-system $D_i \subset M_c$ corresponding to $\psi_i(z, t) = a_i$. But in contrast to the φ_k which define a toroid (connected, compact), D_i is more discontinuous and scattered on M_c as the system departs further from the corresponding integrable system.

Let us then consider the ergodic propositions corresponding to $\tilde{L}_z^w$. To do so let us decompose $\psi_j^{-1}(\delta_i^j(a_j^{-1}))$ into open functions $\Delta_i^j \subset D_i^0$

$$\Delta_i^j = \bigcap_{j \neq i} \psi_j^{-1}(\delta_i^j(a_j)) \cap D_i^0$$

$$Q_{(a_j)}^{(A_j)} \rightsquigarrow \lim_{T \to \infty} \frac{1}{T} \int_0^T U_t \psi_j(\Delta_i^j, t_0)\, \mathrm{d}t =$$

$$= \lim_{T \to \infty} \int_0^T \psi_j(U_t \Delta_i^j, t)\, \mathrm{d}t = \hat{\psi}_j(\Delta_i^j).$$

Now $\hat{\psi}_j$ must be U_t invariant, but $\bigcup_{t_0=t}^{\infty}(U_t \Delta_i^j) = D_i$. $\hat{\psi}_j$ is therefore constant on D_i. It is a function of a_i.

On the other hand the measure $\rho_i(P_t)$ becomes:

$$\lim_{T \to \infty} \frac{1}{T} \int_0^T \gamma_i(P_t)\, \mu(\mathrm{d}P_t)\, \mathrm{d}t = \lim_{T \to \infty} \frac{1}{T} \int_0^T \gamma_i(P_t)\, \mathrm{d}\mu = \hat{\gamma}_i(a_i)$$

since μ and γ_i are U_t invariant $\hat{\rho}_i = \hat{\gamma}_i(a_i)\, \mu$.

The lattice is then composed of a family of Boolean lattices, isomorphic to $\tilde{L}_{\varphi_i}$ defined above. We have a lattice by experimental arrangement (preparing a_i). This is not surprising if we remember that the space state (of the solutions) of a system is defined only by the couple, initial conditions and evolution equations.

To an observable A_j there will be associated a measure of $(\mathbb{R}_{A_j})$ in $L_{\varphi_{A_i}}$ defined by $\hat{\psi}_j(a_i)$.

(b) *Case of Ergodic Systems* [18a]

This is the case where M_c is metric-indecomposable, that is if every invariant measurable set has measure 0 or 1. Then in the decomposition (Section 6 (a)), M_c is the maximum sub-variety of Γ, U_t invariant, and the only observables U_t invariant on M_c are associated with constant functions. We usually express this fact by saying that the trajectory is dense in phase space.

In the Hilbertian formalism [22, 20a] of classical mechanics we then have:

$$U_t = \int_0^{2\pi} e^{i\lambda t}\, \mathrm{d}E_\lambda \qquad E_\lambda\text{: spectral distribution of } U_t$$

$$(\langle f|\, U_t\, |g\rangle = \int_0^{2\pi} e^{i\lambda t}\, \mathrm{d}_\lambda \langle f|\, E_\lambda\, |g\rangle) \tag{6}$$

$$E_{+0} = |k\rangle\, \langle k| \quad \text{with} \quad k(\omega) \equiv \mu(M_c)^{-1/2}, \quad \omega \in \Gamma, \quad \langle k\,|\,k\rangle = 1. \tag{7}$$

Then E_λ has no discontinuity in $\lambda = 0$ for $f \in L^2(M_c, \mu)$:

$$\lim_{\lambda \to +0} \|(E_\lambda - |k\rangle\, \langle k|)\, f\| = 0 \tag{8}$$

and $E_{+0} f = |k\rangle\, \langle k\,|\, f\rangle$ which is the definition of an observable described by f, and ergodic ([4], p. 16).

The observables of a system for which (1, 2) exists are ergodic. The unitary of U_t gives, ($f_\lambda(P)$ being an U_t eigen-function)

$$U_t\{f, g\} = \{U_t f, U_t g\}$$

$\forall P \in M_c\, |f_\lambda(P)| \not\equiv 0$, U_t invariant, and therefore constant (positive) and $\forall f_\lambda \leftrightarrow |\lambda\rangle = |k\rangle\, e^{i\theta_\lambda} (\theta_\lambda(P)$ real) and [23] the proper values λ cannot be degenerate; $\theta_{\lambda_n} \neq n\theta_{\lambda_0} (\lambda_0 = \inf._{0.2\pi} \lambda)$, if not the system is evidently equivalent to a set of harmonic oscillators. A sufficient condition for this not to take place is the existence of λ_1, λ_2 proper frequencies with

$$m\lambda_1 + n\lambda_2 = 0 \qquad m, n \in N \Rightarrow m = n = 0 \tag{9}$$

$\{\lambda_i\}$ is then dense in $[0, 2\pi]$ and $\forall f \in L^2(M_c, \mu)$:

$$|f\rangle = \sum_i C_{\lambda_i} |\lambda_i\rangle. \tag{10}$$

LEMMA. *For all observables $f \in L^2(M_c, \mu)$ the correlation function*

$$r(n) = \frac{|\langle f|\, U^n\, |f\rangle|}{\langle f\,|\, f\rangle} \tag{11}$$

converge in mean (discrete measure) towards zero.

In fact $U_t^n |f\rangle \equiv |f_n\rangle$ considered as a random event on (M_c, μ) is a stochastic process with $r(0)=1$, $r(n) \leqslant 1$, but starting from the spectral resolution of U_t we have the representation of Cramer-Khinchin type:

$$r(n) = \int_0^{2\pi} e^{i\lambda n} \, \mathrm{d}_\lambda \frac{\langle f | E_\lambda | f \rangle}{\langle f | f \rangle} = \int_0^{2\pi} e^{i\lambda n} \, \mathrm{d}F(\lambda) \tag{12}$$

which with Equation (8) gives:

$$\lim_{N \to \infty} \frac{1}{N} \sum_{n=0}^{N-1} r(n) = 0. \tag{13}$$

Consequences.

For all s_i (Section 6, Equation (7)) if $\lim_{t \to \infty} \rho_i(t)$ exists:

$$\lim_{t \to \infty} \rho_i(t) = k\mu \quad \text{with} \quad \mu \text{ as in Section 6 (Equation (5)).} \tag{14}$$

In fact $\rho_i(t) = \gamma_i(P_t) \, \mu(\mu$ continuity of $\rho_i)$. If χ_A designated the characteristic function of a $A \subset M_c$ which is measurable, we have:

$$\forall \chi_A; \; m^\infty(A) = \lim_T \frac{1}{T} \int_0^T (U_{-t}\gamma_i, \chi_A) \, \mathrm{d}t = (\hat{\gamma}_i, \chi_A)$$

and

$$\rho_i^\infty = \hat{\gamma}_i \mu \quad \text{and} \quad \rho_i^\infty \text{ is } U_t \text{ invariant}.$$

Now for U_t ergodic and irreversible, there exists only one invariant measure [20b], therefore $\gamma_i = k \cdot (p \cdot p)$ and $k \neq 0$.

THEOREM. *The lattice of propositions of the system Σ for the ergodic model is Boolean atomic.*

Let $A \in \mathcal{O}_\Sigma$, $f_A(\varphi, \psi_i)$ the associated Borel functions:

$$Q_E^{\hat{A}} = \left\{ P \in \Gamma, \lim_{T \to \infty} \frac{1}{T} \int_0^T f_A \, \mathrm{d}t \in E \right\} =$$

$$= \{ P \in \Gamma, E_{+0} f_A \in E \} = \{ c_i, \tilde{f}(c_i) \in E \}$$

with

$$\tilde{f}(c_i) = \int_{M_c} f_A(P) \, k \, \mathrm{d}\mu.$$

Where $\tilde{f}(c_i)$ are m measurable and $\varphi_i(P)$ uniform and continuous on Γ, $\tilde{f}$ is a Borel function on $\mathbb{R}^m$.

8. The Stochastic Model

By retaining all the postulates, if we wish further to consider as physical propositions those which concern the observables defines on M_c, we obtain the stochastic model. In fact, any function on M_c is a possible random variable and the family induced by U_t is then a random function. Two possibilities then present themselves: either the observations are defined at one precise instant, or the observations concern the over-all values of the observables over a definite interval in time. We will call the first 'pure stochastic models' and the second Brownian (or Gaussian).

In every case, we have at the start on M_c a distribution ρ, U_t invariant and stationary established as the result of the ergodic process. The variables $f(\omega)$ are not correlated with the $f(\omega_0)$ (cf. Section 7, Equation (13)). The observation considered in this model takes account of compatible times (sufficiently large) with the establishment of the ergodic state of equilibrium.

(a) *The Purely Stochastic Model*

We consider a sequence of measurements at times $t_0, \ldots, t_l$ of the observables A_i $i=1, k=2(n-m)$, independent, defined on M_c and characterising completely the system with the constants of motion $\varphi_i = c_i$; $t_{j+1} - t_j \gg T$ where T is the ergodic equilibrium time. This sequence concernes a set of identical system (repeatability) and provides a frequency distribution $\nu(k_i)$ of the k results obtained at the time of the ith measurements.

For the series t_i the lattice of propositions relative to M_c is then that associated with the σ algebra generated by the direct l product of the σ algebras of the Boolean parts of $D^k \subset \mathbb{R}^k$ where D^k is the set of the results of the $\nu(k)$ measurements. On each of these lattices we have a measure of probability corresponding to the $\nu(k)$. As in (V) this lattice is ortho-isomorphic with a sub-lattice of a modular reducible lattice which can here be constructed directly.

Let us put:

$$\nu(k_i^\sigma) = \bar{\alpha}_i^\sigma \alpha_i^\sigma \qquad k_i^\sigma \in \mathscr{B}_i(D^{k_i}), \tag{1}$$

where the phase of the α_i^σ is indeterminate and $\mathscr{B}_i(D^{k_i})$ is a sub-base of the σ algebra D^{k_i} and:

$$\sum_{\sigma \in \mathscr{B}_i} v(R_i^\sigma) = 1 \leftrightarrow \{\alpha_i^\sigma\}_\sigma \sim l^2.$$

With each result k_i^σ we associate $|\psi_i\rangle$ with:

$$v(k_i^\sigma) = |\langle \gamma_i^\sigma \mid \psi_i \rangle| 2, \tag{2}$$

where the $|\gamma_i^\sigma\rangle$ are the vectors base of $\mathscr{H}_i$, separable and isomorphic to l^2. For each value system of the φ_i we therefore have a base of l^2 for which the probabilities $v(k_i^\sigma)$ are the traces of $|\psi_i\rangle$ on that base. The implications are defined by the elements of the transition matrix:

$$[T(i,j)]_{\sigma_i \sigma_j} = v(k_j^{\sigma_j} \to k_i^{\sigma_i}) \geqslant 0, \tag{3}$$

where v is in this case the transition probability:

$$v(k_j^{\sigma_j} \to k_i^{\sigma_i}) = \frac{v(k_i^\sigma, k_j^\sigma)}{v(k_j^\sigma)} \qquad \begin{array}{l} i > j \text{ deduction} \\ i < j \text{ induction} \end{array}$$

and:

$$\sum_{\sigma_i \in \mathscr{B}_i} [T(i,j)]_{\sigma_i \sigma_j} = 1 \tag{4}$$

then by making use [10] of the indetermination of phase of the α_i we have:

$$[T(i,j)]_{\sigma_i \sigma_j} = \left| \frac{\langle \gamma_i^{\sigma_i} | \mathscr{T}(ij) | \gamma_j^{\sigma_j} \rangle}{\| \mathscr{T}(ij) \mid \gamma_j^{\sigma_j} \|} \right|^2. \tag{5}$$

The operator $\mathscr{T}(i,j)$ defined by (5) is the restriction to M_c of the operator of evolution of the system. It is isomeric as a result of (4). Its stationary nature implies further that $\mathscr{T}(i,j) = \mathscr{T}(i-j)$ and its unitary nature that the process is causal. In fact with (1), (3) and (5) we have:

$$\forall i > j \; \forall |\psi_j\rangle \leftrightarrow \{\alpha_j^\sigma\} \quad \text{and} \quad \forall \mathscr{H}^i \text{ of base} \quad \{|\gamma_i^{\sigma_i}\rangle\}_{\sigma \in \mathscr{B}_i}$$

$$|\gamma_i^{\sigma_i} \mathscr{T}(ij) |\psi_j\rangle|^2 = \sum_{\sigma_j} \langle \gamma_i^{\sigma_i} | \mathscr{T}(ij) | \gamma_j^{\sigma_j} \rangle \langle \gamma_j^{\sigma_j} \mid \psi_j \rangle|^2$$

$$0 = \langle \gamma_i^{\sigma_i} | \sum_{\sigma_j} \mathscr{T}(ij) | \gamma_j^{\sigma_j} \rangle \langle \gamma_j^{\sigma_j} \mid \psi_j \rangle [\langle \psi_j | \mathscr{T}^+(ij) | \gamma_i^{\sigma_i} \rangle -$$

$$- \langle \psi_j \mid \gamma_j^{\sigma_j} \rangle \langle \gamma_j^{\sigma_j} | \mathscr{T}^+ | \gamma_i^{\sigma_i} \rangle].$$

Since the base $|\gamma_i^{\sigma_i}\rangle$ is complete in $\mathscr{H}_i$, $\mathscr{T}(ij)$ unitary implying that the ensemble $\mathscr{T}(ij)\,|\gamma_j^{\sigma_j}\rangle$ is free, the inequality only applied when:

$$\begin{cases}\langle\gamma_j^{\sigma_j}\mid\psi_j\rangle=0\\ \langle\gamma_i^{\sigma_i}|\,\mathscr{T}(ij)\,|\psi_j\rangle=\langle\gamma_i^{\sigma_i}|\,\mathscr{T}\,|\gamma_j^{\sigma_j}\rangle\,\langle\gamma_j^{\sigma_j}\mid\psi_j\rangle\end{cases}$$

$\forall|\psi_i\rangle$ and $\forall\sigma_i$. Then:

$$|\psi_j\rangle=|\gamma_j^{\sigma_j}\rangle\,\langle\gamma_j^{\sigma_j}\mid\psi_j\rangle \tag{6}$$

for one single σ_j.

The state $|\psi_i\rangle=\mathscr{T}(ij)\,|\psi_j\rangle$ can therefore only have as its origin the state $|\gamma_j^{\sigma_j}\rangle$ obtained for the preceding observation j. This formalism is similar to that of quantum mechanics, for observables with denumberable spectrum. Inversely, it is established that the quantum model of experimental observables is a stationary unitary restriction of the pure stochastic model. This method could appear to be a good means of making a model with hidden variables.

All kinetic models are connected with this model. The observations have a continuous spectrum and the implications are provided not by (5) but by the solutions of a kinetic equation [8]. All these equations introduce (4) explicitly for $i<j$ and $j>i$, that is a unitary operator $\mathscr{T}(ij)$ introducing the different stochastic models of the wave formalism of quantum mechanics.

(b) *The Brownian Model*

The propositions concern mathematical expectations and mean-square deviation of the observables on M_c provided with the invariant measure by which the random variables are of Markovian and Gaussian type. This hypothesis is justified for C systems (cf. Arnold Avez, *op. cit.*, Chapter III). On the other hand, for linear mechanical systems which are symmetrical by translation, this hypothesis is probably true in the limit of $n\to\infty$ degrees of freedom. Physically the hypothesis of strong mixing is imposed on the system so that it will provide statistically regular events [20b]. This implies the presence of the observing equipment. The principle of repetition (reproducibility) of the observations (Section 3(c)) provides the Markovian character. Lastly, the Gaussian character is imposed when we limit ourselves to two parameters the mean and the standard deviation. It is then known [11] that there exists only one equivalent Gaussian stochastic process. Reciprocally, any Gaussian-Markovian process is strongly mixing; to see this we consider a base of $L^2(M_c, \mu)$ defined by:

$$U_t^n \,|\gamma\rangle, \langle z \mid \gamma\rangle = e^{i\gamma z},$$

where z is the Gaussian variable defined over M_c.

Then [40]

$$\lim_{n\to\infty} \langle\gamma'|\, U^n \,|\gamma\rangle = \lim_{n\to\infty} e^{-(\gamma'^2+\gamma^2)/2}\, e^{\gamma\gamma' r(n)} = \langle\gamma' \mid k\rangle \langle k \mid \gamma\rangle$$

by using Equations (13) and (8) of Section 7.

In this case the construction of the Hilbert space is well known [41]. To the stochastic variable $z(t, \alpha)$ of which $\Delta z(t_2, t_1, \alpha)$ are Gaussian, centred and of variance $|t_2 - t_1|$, we associate for all observables $f(P_t)$ the stochastic integral

$$F(\alpha) = \int_{-\infty}^{\infty} f(t)\, \Delta z(t, \alpha)$$

(in the sence of mean square convergence) with

$$\langle|F(\alpha)|^2\rangle = \int_T |f(t)|^2 \,\mathrm{d}t \qquad T = [0, 1].$$

Then for all orthonormal bases of $L^2[0, 1]$ of function of t there corresponds a complete system of centred Gaussian functions $\Phi(\alpha)$ which are statistically independent.

For all systems of observables $\varphi_i = c_i$ there corresponds a family $f_k(P)$, $P \in M_c$ which is orthonomal, of eigen functions of U_t on M_c base of $L^2(M_c, \mu)$ and to every observable $Q(P)$ of the system there corresponds a Brownian stochastic variable:

$$Q(\alpha) = \int_{M_c} Q(P_t)\, \mathrm{d}z(P_t, \alpha) = \sum_N^{\infty} A_N \psi_N(\alpha). \tag{2}$$

The second equality in the sense of mean-square convergence on $L^2([0, 1], d\alpha)$ where $\psi_N(\alpha)$ is the stochastic base [7] associated with the f_k.

The predictions concerning $(1/T) \int_0^T Q(P_t)\, \mathrm{d}t$ are given by the probabilities of $Q(\alpha)$, $\alpha \in [0, 1]$, that is by the known distributions of the $\psi_N(\alpha)$'s and of the coefficients A_N. In order to determine the N's for which the A_N's are non-zero, we consider the development:

$$Q = \sum_k c_k f_k \qquad \sum_k |c_k|^2 = 1 \tag{3}$$

and the theorem [42]: let $\chi(z)$ and $\psi(z)$ be two orthogonal functions, within $L^2(M_c, \mu)$; then:

$$\frac{\|\chi\|^2}{\|\chi\|^2 + \|\psi\|^2} = m\{\alpha; |\Xi(\alpha)| \geqslant |\psi(\alpha)|\}, \tag{4}$$

where:

$$\Xi(\alpha) = \int_{M_c} \chi(P)\, \mathrm{d}z(P, \alpha).$$

Thus we have associated the probability defined by (4) with the proposition $k \in \mathrm{Sp}(Q)$.

We can therefore characterise the lattice of propositions of the Brownian model as the direct product of the lattice of the sub-spaces of the Hilbert space of base ψ_N or f_k (of continuous dimension as soon as the system becomes strongly mixing) and the Boolean lattice of the parts of $\mathbb{R}^m$ where m is the number of controllable first integrals (constant in the preparation of the Brownian system).

9. The Model of Statistical Mechanics

All the hypotheses (contained in the chapters before Section 8) being retained, let us now consider the influence of the process of preparation. This still necessitates the presence of a macroscopic apparatus Σ' with phase space Γ' and which possesses observables φ_i' of the same type as those of the prepared system. The observables (invariant integrals) of the total system $\Sigma \cup \Sigma'$ are:

$$\Phi_i = \varphi_i' + \varphi_i \qquad i = 1, m.$$

Postulate M. The Interaction Between Σ' and Σ is Weak

We make use of the usual expression [27] of this: for all choices p' and p of absolutely continuous measures of μ' and μ on Γ' and Γ, there exists an

m-set $\gamma=(\gamma_1,\ldots,\gamma_m)$ of values of Φ_i such that:

$$p'\{P'\in\Gamma',\ \alpha_i<\varphi_i'(P')+\varphi_i(P)-\gamma_i<\beta_i\}=k \tag{1}$$

k is independent of $P\in\Gamma$.

Then:

$$\forall f\in L^1(\Gamma,p)\qquad \lim_{T\to\infty}\frac{1}{T}\int_0^T f(U_tP)\,\mathrm{d}t=\int_\Gamma f(P)\,\mathrm{d}p \tag{2}$$

for almost all P of Γ such that:

$$\varphi_i(P)={}^1\gamma_i\quad (i=1,m)\qquad \gamma_i={}^1\gamma_i+\gamma_i'.$$

On the other hand p and p' must be invariant by U_t and U_t'; then there exists (θ_i), $i=1$, m constant with:

$$p=a_\theta^{-1}\,e^{-\theta\cdot\varphi}\mu\qquad\text{and}\qquad p'=b_\theta^{-1}e^{-\theta\cdot\varphi'}\mu' \tag{3}$$

a_θ and b_θ being the normalization factors:

$$a_\theta=\int_\Gamma e^{-\theta\cdot\varphi}\,\mathrm{d}\mu\qquad b_\theta=\int_\Gamma e^{-\theta\cdot\varphi'}\,\mathrm{d}\mu' \tag{4}$$

then for every θ we have a set of values γ_θ of first integrals of Σ and Σ' such that:

$$\forall f\in L^1(\Gamma,e^{-\theta\cdot\varphi}\mu),$$

$$\lim_{T\to\infty}\frac{1}{T}\int_0^T f(P_t)\,\mathrm{d}t=\frac{1}{a_\theta}\int_\Gamma f(P)\,e^{-\theta\cdot\varphi(P)}\,\mathrm{d}\mu \tag{5}$$

almost everywhere within $\{P,\ \varphi_i(P)=(\gamma_\theta)_i\quad i=1,m\}$; the values $(\gamma_\theta)_i$ are determined by the measurements (time averages) of the invariant of the system Σ: $\hat{\varphi}=c=\langle\varphi\rangle_\theta=\gamma_\theta$.

It is then well known that all the predictions provided by this model are obtained starting from a_θ, the partition function for m uniform, time independent integrals. If we denote by α the (intensive) variables to which Σ is subjected, and which therefore depend on the first integrals (without restriction of the problem, this dependence is C^∞) we have:

$$\begin{cases} \gamma_i = -\dfrac{\partial \log a_\theta(\alpha)}{\partial \theta_i} \quad \text{expected value of } \varphi_i \\ \sum\limits_i \theta_i \cdot X_{ij} = -\dfrac{\partial \log a_\theta(\alpha)}{\partial \alpha_j} \end{cases} \tag{6}$$

X_{ij} characterizes the influence of α_j on φ_i (extensive forces or variables). $G = \log a_\theta + \theta \cdot \gamma$ playing the part of entropy and θ^{-1} that of the generalized temperature.

$$c_j^i = \frac{\partial^2 \log a}{\partial \theta_i \partial \theta_j} = -\frac{\partial \gamma_i}{\partial \theta_j}$$

are the specific heats with constant α.

The introduction of the macroscopic apparatus Σ' has therefore had as a consequence that the values of the first integrals φ_i of Σ become random, and that for each state an m-set of values is introduced, θ characterizing the preparation. θ then becomes an ensemble of observables on the same foot as the φ_i and for Σ, φ and θ are the random variables of mean values γ, Θ where Θ is the value of θ which satisfies (6).

THEOREM. *The observables* (θ_i, φ_i) $i = 1, m$ *are complementary observables (in the quantum sense).*

The probability density on Γ being $f(z) = a_\theta^{-1} e^{-\theta\varphi}$, we know [27] that:

$$\langle \theta_i \varphi_k \rangle = \delta_{ik} \tag{7}$$

that is the random variables are independent.

For $(\theta_i; \varphi_i)$ let us consider the variations of (6) for $\alpha_i = Cte$, $i \neq j$.

$$\alpha_j \text{ variable and } \begin{cases} \theta_k = Cte \\ \varphi_k = Cte \end{cases} \quad k \neq i \text{ putting } \Theta_1 = \frac{1}{kT} \, (U_1 = E)$$

we have

$$\begin{aligned} &\langle \Delta\varphi_i \rangle^2 = \left(\frac{\Theta_i}{\Theta_1}\right)^2 c_{ij} \qquad \langle \Delta\theta_i \rangle^2 = \left(\frac{\Theta_i}{\Theta_1}\right)^2 c_{ij}^{-1} \\ &\langle \Delta\varphi_i \rangle^{1/2} \, \langle \Delta\theta_i \rangle^{1/2} = \left(\frac{\Theta_i}{\Theta_1}\right)^2 \qquad \forall j = 1, m. \end{aligned} \tag{8}$$

THEOREM. *The lattice of propositions of statistical mechanics is* *modular. To each set A_i $i=1, m$ of observables chosen within* (θ, φ) *there corresponds a state where these observables are predicted and where the probabilities of the values of the others are given by:*

$$F(\varphi, \theta)=\frac{e^{-\theta\cdot\varphi}\Omega(\varphi)}{a_\theta}.$$

We are in the same position as case (Section 6).

It may be remarked that for each one of these m-set we have associated one pure phase, that is an ergodic measure (Gibbs' phase rule).

In conclusion, the presence of the macroscopic equipment Σ' has modified the Boolean ergodic lattice of Σ into a modular lattice. As a result of (1) no passage to the limit is possible. It is noted that (1) is also valid in quantum mechanics [2] and that the inequality relations (8) are of the same type as those of Heisenberg.

10. THE THERMODYNAMIC MODEL

POSTULATE (T): *the only observable quantities are functions of the conjugate variables* (γ_θ, Θ), *the intensive variables* α *remaining the experimental constraints which define* Σ.

An immediate consequence will be the absence of time in the set of thermodynamic propositions. In order for (T) to be admissible, following the previous postulates, two possibilities offer themselves:

(a) The fluctuations of the quantities γ_θ and θ are negligible (with a meaning to be stated)

(b) There exists an experimental model for which the description of Σ only contains the mean values of (γ_θ, θ). The mean-square deviations are redundant if they are not zero.

Examination of these two conditions, their equivalence and their implications will provide the lattice of propositions sought for. We will see that there exists an equivalent experimental lattice of propositions (Section 11).

A. *Conditions of Boltzmann Type*

In stochastic or Brownian models, the fluctuations are negligible if:

$$T_{ij}\simeq\delta_{ik} \quad \text{or} \quad Z(t, \alpha)\simeq\xi(t)\,\mathrm{d}\alpha \tag{1}$$

that is if the Markov process is quasi-determinist in the sense of (Section 9 (Equation (6)). Now if a Markov mechanical system is finite, no transition state exists. Further Section 9 (Equations (3) and (4)) give:

$$T_{ij} \leqslant \exp\left[-\Delta_{ij} \int_{\Gamma} e^{-\theta\varphi} \, d\mu\right], \tag{2}$$

where i and j are two states of Σ for the same ergodic state of $\Sigma \cup \Sigma'$. We must therefore impose a condition that a_θ should satisfy (1). No general results exist. In the case where:

$$\theta\varphi = \sum_{i=1}^{\nu} \theta_i \varphi_i$$

that is to say in the case of 'sum functions', Schwartz' inequality applied to Section 7 (Equation (4)) and Tchebicheff's inequality for all $g \in L_2(\Gamma, \mu)$ with:

$$g = \sum_{k=1}^{\nu} f_k \tag{3}$$

give:

$$\frac{\langle (g-\hat{g})^2 \rangle}{\langle g^2 \rangle} = 0\left(\frac{1}{\nu}\right)$$

that is the condition $N \to \infty$ necessary in (a) is then sufficient to make the fluctuations negligible. In the cases where (3) is not verified, the limit of $\log a_\theta$ is to be studied directly. For all potentials χ such that the energy of $X \subset \Gamma$

$$U(X) = \sum_{z \in X} \chi(z_i)$$

is bounded [37] (in particular for a two-body interaction [15] or a short range interaction) we have the limit:

$$\lim_{N \to \infty} \frac{1}{N} \log a_\theta(N) = s(\theta, \lambda)$$

where:

$$\lambda = \lim_{N \to \infty} \frac{\varphi(\lambda, N)}{N}$$

N is the number of repeats of Γ and $Ns(\theta, \lambda) = \log a_\theta$. Then:

$$\langle \varphi_\theta(\lambda) \rangle = \gamma_\theta \qquad \langle \varphi_\theta^2(\lambda) \rangle = 0$$

and all the quantities defined in Section 8 (Equation (6)) are well defined.

Consequences

The fact that it is the function $\log a_\theta$ and not a_θ which defines thermodynamic magnitudes, introduces a supplementary equivalence relationship into S_Σ. For example, two different system of n repeats of Σ, in the same state, provide the same $s(\theta, \lambda)$ although

$$a_\theta(N)/a_\theta(N-n) \underset{N\to\infty}{\to} 0.$$

The set of first integrals for which the value of $s(\theta, \lambda)$ is not affected will be known as the mechanical components of Σ; for these observables s is constant although the states are different. It will be noted q_i $i = 1, m$ the remaining macroscopic quantities:

$$s = s(\theta_i q_i) \qquad (i = 1, m. \tag{4}$$

Finally:

$$\lim_{N\to\infty} \frac{1}{N} \log a_\theta(n) = s(\theta_i q_i)$$

is subadditive of (q_i) and is convex in θ_i. Gibbs' phase rule is then valid.

B. *Statistical Conditions*

Let us consider:

$$p(\theta z) = F(\theta, \varphi(z) = \frac{e^{-\theta\varphi}}{\alpha_\theta} \Omega(\varphi) \tag{5}$$

as defined in Section 8 (Equation (10)) and let us put:

$$\mathrm{d}P(z/\theta) = \frac{e^{-\theta\varphi(z)} \Omega(\varphi)\, \mathrm{d}\varphi\, \mathrm{d}\theta}{p(\theta)\, \mathrm{d}\theta} \tag{6}$$

the probability for Σ to be in a state prepared by θ and defined by z then (5) is simply the probability of observing $\varphi(z)$ when θ is certain.

In general such a probability is written as:

$$p(z/\theta)=\frac{\pi(z,\theta)}{P(\theta)}.$$

THEOREM [23]. *If $p(z/\theta)$ is analytic and non-zero in an open set $0\subset\Gamma\otimes\mathbb{R}^m$ and if $\varphi_1,\dots,\varphi_\nu$ are continuous on $\mathbb{R}$ and $m<\nu$; the necessary condition for $\varphi_1,\dots,\varphi_\nu$ to form a sufficient statistic for p is that for all $(\alpha,\beta)\in 0$ in the neighbourhood of this point:*

$$p(z/\theta)=\exp\left[-\sum_{k=1}^{\mu}\Theta_k X_k+\Theta_0+X_0\right], \tag{7}$$

where Θ_j are uniform and analytic in θ and X_i in z is the neighbourhood of (α,β). If m is the smallest value for which (7) *is valid then:*

$$\sum_{i=1}^{m} X_k(z_i)=V_k(\varphi_i,\dots,\varphi_\nu),$$

where V_k is uniform.

In other words, if we can reproduce Σ in the state S_Σ characterized by θ, then $p(\theta,z)$ is given by (5) and reciprocally (5) is the necessary form of the law of probability such that the statistics concerning $\varphi(z)$ shall lead to good estimates of the θ. Equations (6) of Section 8 then concern fully defined quantities. If for example we impose the following conditions on the φ:

$$\varphi_i(z,\lambda)\in[\gamma_i-f(\beta),\gamma_i+f(\beta)] \quad \theta \text{ fixed (external coupling)}$$

then:

$$F(\varphi,\theta,\lambda)=\int_{\Gamma_\theta}\mathrm{d}P(z/\theta)=$$

$$=\frac{1}{B}\exp\left[-\sum_{n=1}^{m}\varphi_n\theta_n\right]\int_{\Gamma_\theta}\mathrm{d}z=\frac{1}{a_\theta}\exp\left[-\sum_{n=1}^{m}\varphi_n\theta_n\right]$$

where

$$a_\theta=\int_{\Gamma}e^{-\beta E(z)}\,\mathrm{d}z$$

and:

$$\mathrm{d}P(z/\theta\beta)=\frac{e^{-\beta E(z)}}{a_\theta}\,\mathrm{d}z. \tag{8}$$

Then (maximum likelihood), the solutions of:

$$\frac{\partial F}{\partial \varphi_n} = 0 \qquad \theta_n = \frac{\partial \log a_\theta}{\partial \varphi_n} \tag{9}$$

are the most probable values of (φ, θ). These are the Equations (6) of Section 8.

C. *Thermodynamic Propositions*

We have associated an ensemble of m observables q_i and a function $s(\theta_i q_i)$ with every system Σ. To every state of equilibrium there corresponds a point of $\mathbb{R}^m$ and a value of s, the vector tangent to s at this point defining θ. This is an exact description of classical mechanics [28] where the phase space $\mathscr{M}_{V*}$ (linear form on the tangent space of the configuration space) is replaced by S, for the co-ordinates $\theta_i = \partial/\partial q_i$ and $\pi(\theta_i) = q_i$). The equivalent descriptions [25] are given by the $F(A_i, B_i)$ where (A_i, B_i) is deduced from $(\theta_i q_i)$ by a product of Legendre transformations and F the corresponding transformations of S. Since the additive and convexity properties are conserved, we deduce at once that the lattice of propositions of a thermodynamic system in equilibrium is the lattice of the Borelians of $\mathbb{D} \subset \mathbb{R}^{m-1}$.

$\mathbb{D}$ is a compact part of $\mathbb{R}^{m-1}$. A system of co-ordinates is known as a system of variable state for Σ and the equation of $\mathbb{D}$ in $\mathbb{R}^m$ is the equation of state. We then have:

THEOREM. *The lattice of thermodynamic propositions is Boolean complete and atomic.*

11. Experimental Thermodynamic Propositions

Starting from the considerations of Section 3 we will now show that experimental thermodynamic conditions provide a lattice of experimental propositions isomorphic with that of the model of Section 10. We are concerned with a total description and in conformity with the theorem of Section 4, the lattices are Boolean [6b].

We can characterize 'thermodynamic' experiments relative to $\mathscr{A}$, $\mathscr{S}$, S defined in Section 3.

(1) $\mathscr{A}_T$ the complete ensemble of measuring instruments has no ele-

ment equivalent to a clock, but we suppose that we are able to say whether one experiment takes place before or after an other (the notion of succession is acquired).

(2) $\mathscr{S}_T$ the same complete ensemble is used for Σ or for $\bigcup_n \Sigma_i$ for all finite n and $\forall i$, $\Sigma_i = \Sigma$; in particular, the components of $\bigcup_{i=1}^n \Sigma_i^i$ are discernable.

(3) S_T the states are dispersion free. L and Σ_1 (resp. Σ_2) be a system of $\mathscr{S}_T$ prepared in the state s_1 (resp s_2). The system $\Sigma = \Sigma_1 \cup \Sigma_2$ is a system for which $\mathcal{O}_\Sigma = \mathcal{O}_{\Sigma_1} = \mathcal{O}_{\Sigma_2}$. $\mathscr{L}_\Sigma$ is the free product of $\mathscr{L}_{\Sigma_1}$ by $\mathscr{L}_{\Sigma_2}$. For every element of $\mathscr{L}_\Sigma$, $Q_E^A = Q_{E_1}^A \wedge Q_{E_2}^A$, we define $sQ_E^A = s_1 Q_{E_1}^A \times s_2 Q_{E_2}^A$ and we note that $s = s_1 + s_2$.

THEOREM. *For a thermodynamic system, the ensemble S_T of the thermodynamic states provided with + is a commutative monoid.*

The definition in Section 3 (c) of the states and the complete property S_T lead to a simplification law for +. In fact if $\forall Q_1 \in \mathscr{L}_{\Sigma_1}$, $\forall Q_2 \in \mathscr{L}_{\Sigma_2}$ $\forall Q_3 \in \mathscr{L}_{\Sigma_3}$ and $s_1 + s_3 = s_2 + s_3$ then:

$$s_1 Q_{E_1}^A = s_2 Q_{E_2}^A \mapsto s_1 = s_2.$$

DEFINITIONS. *We define a thermodynamic transformation as any ordered pair of states.*

This is called natural if the states correspond to two successive recorded observations ($s_1 > s_2$); then ($s_1 < s_2$) will be called antinatural and (s, s) nul.

It is called reversible if it is natural and anti-natural.

The interpretation of the notion of a natural transformation leads us to make the following postulate:

$$(s_1 > s_2) \text{ and } (s_1 > s_3) \mapsto (s_2 > s_3) \text{ or } (s_3 > s_2).$$

The set of natural transformations from the same origin is simply ordered.

Group of Transformations

By putting:

$$(s_1, s_2) + (s_1' + s_2') = (s_1 + s_1', s_2 + s_2') \tag{4}$$

$$-(s_1, s_2) = (s_2, s_1) \tag{5}$$

we have a group structure T. If in addition we consider the group quotient by the equivalence relation

$$(s_1, s_2) \sim (s_1 + a, s_2 + a) \tag{6}$$

$T/T_0 \sim \mathcal{T}$ is a partially ordered group, without any nilpotent element and:

$$(s_1, s_2) > (s_1', s_2') \leftrightarrow (s_1 + s_2' s_2 + s_1') \tag{7}$$

– the natural transformations $\mathcal{T}_N$ being the positive cone of $\mathcal{T}$
– two notable sub-groups of $\mathcal{T}$ are $(\mathcal{T}_N) \vee (-\mathcal{T}_N) \equiv \mathcal{T}_P$ and $(\mathcal{T}_N) \wedge (-\mathcal{T}_N) \equiv \mathcal{T}_R$:

we have a family of homomorphisms of S in $\mathcal{T}$ by

$$s \in S \mapsto \sigma = (ks, ls), \text{ if } k > 1, \text{ positive integer, isotonic} \\ \text{if } k < 1, \text{ positive integer, antisotonic.} \tag{8}$$

Thermodynamic Space

By experimental definition this contains all the non-equivalent transformations where this is due to reversibility, that is:

$$\mathcal{G} = \mathcal{T} / \mathcal{T}_R.$$

Thermomechanical Space

The elements of this space characterize all possible transformations:

$$\mathcal{Q} = \mathcal{T} / \mathcal{T}_R.$$

Note. By (8) we define a natural homomorphism of S on $\mathcal{G}$ and on $\mathcal{Q}$:

$$(s_1, s_2) \in \mathcal{T}_R \rightarrow \sigma_{\mathcal{G}}(s_1) = \sigma_{\mathcal{G}}(s_2)$$
$$(s_1, s_2) \in \mathcal{T}_P \rightarrow \sigma_{\mathcal{Q}}(s_1) = \sigma_{\mathcal{Q}}(s\ ).$$

We have a simple order on a part of S. In order to compare (that is to measure) the states amongst themselves, we must extend this order relation. To do this, we make use of the partial order of $\mathcal{T}$ which is well defined experimentally.

DEFINITION. *$s_1 < s_2$ for $s_1 \in S$, and $s_2 \in S \leftrightarrow \exists n$, positive integer and $c \in S$ where*

$$(ns_1 + c, ns_2) \in \mathcal{T}_P. \tag{10}$$

The fact symbolized by (10) is a follows: one state is smaller than another, if we can reach it starting from the second by the mean of a possible operation.

In order for thermodynamics to be a quantitative science, we must have S totally ordered by (10). In other words, it is necessary for there to exist a state e, starting from which all the states of the system may be reached, up to an operation (that which consists of adjoining a replica of Σ in a possible state).

(0_1) AXIOM. *$\forall \Sigma \in \mathscr{S}_T \exists e_\Sigma$ and $\forall s \in S$ $\exists n$ positive integer and $s < ne$.*

THEOREM. *There exists a function $\varphi : S_\Sigma \to (\mathbb{R}^+)^n$ which is additive, constant on all states of possiible transformation, strictly positive for non-equivalent states, bounded.*

Proof. Let us consider:

$$f(s_1, s_2) = \inf_{Q+} \left\{ q \in Q^+,\ n_2 s_2 < n_2 s_1 + n_1 e,\ q = \frac{n_1}{n_2} \right\} \tag{11}$$

then:

$$(s_1, s_2) \in \mathscr{T}_P \mapsto f = 0; \qquad f \in [0, \infty[$$
$$f[(s_1, s_2) + (s'_1 s'_2)] \leqslant f(s_1, s_2) + f(s'_1, s'_2)$$
$$f(ns_1, ns_2) = nf(s_1 s_2).$$

We thus have a function sublinear on $\mathscr{T}$, bounded, linear on every subgroup of $\mathscr{T}$ generated by (s_1, s_2) and $\mathscr{T}_P$. We use the Hahn-Banach theorem [12] to extend this function additive on $\mathscr{T}$. By homomorphism [8] we obtain the φ sought for.

Rule of Interpretation

The existence of e_Σ imposed by axiom 0_1 but inaccessible by direct thermodynamic experiment, leads to interpretation of experiments carried out macroscopically (n large) at the microscopic level ($n \sim 1$). In order for the theory to be consistent, it is necessary for a thermodynamic transformation to be significative of the theoretical transformation corresponding to the microscopic level. In other words, if a macroscopic process is possible the microscopic process is possible. This is formally expressed as:

(0_2) AXIOM. *$\forall (s_1, s_2) \in \mathscr{T}$ if $\forall \varepsilon > 0 \exists m$ and n positive integers with $m/n < \varepsilon$ and if:*

$$\forall k_1 \eta_1 < me,\ k_2 \eta_2 < ke;\ \frac{k}{k_i} < m \mapsto (ns_1 + \eta_1 > ns_2 + \eta_2)$$

then: $(s_1 > s_2)$.

Note. The axiom (0_2) expresses the fact that $\mathscr{T}_N$ is closed for the order topology (10):

$$\|(s_1 s_2)\| = \inf_n \left(\frac{\|s\| + \|s'\|}{n} \right) \text{ for } (ns, ns') \sim (s_1 s_2)$$

$$\|s\| = \inf m/R \; ks < me.$$

Then $\mathscr{T}_{-N}$ is closed, $\mathscr{T}_P$ and $\mathscr{T}_R$ are closed, therefore $\mathscr{G}$ and $\mathscr{Q}$ are separable. There also exists a sufficient set of φ_i, continuous, additive, bounded by S in $\mathbb{R}^+$ to separate the points of S. (This is the symbolization of the notion of sufficient statistics as seen above.)

THEOREM. *There exists a function Z: $S_T \to \mathbb{R}$ which is additive, constant on equivalent states by reversibility, strictly increasing on irreversible transformations.*

Proof. Let us consider

$$z(s_1 s_2) = \inf_{Q^+, S_T} \left\{ r \in Q^+, \; x \in S_T, \; (ns_1 + x > ns_2 + me); \frac{m}{n} = r \right\}.$$

The transformation $(x; me)$ is the smallest which makes (s_1, s_2) natural. z is a good quantity for comparison of the irreversibility of transformations with that which terminates by a multiple of e.

We then verify that:

$$z \in [0, \infty[, \; z(-\mathscr{T}_N \cup \mathscr{T}_R) = 0,$$

z is sublinear on $\mathscr{T}$, strictly positive and additive on $\mathscr{T}_N/\mathscr{T}_R$.

In the same way as f we extend to a function F which is positive bounded, and additive on $\mathscr{T}$. By the homomorphism (8) we obtain Z.

Conclusion

To all thermodynamic states we have associated a collection of $m+1$ scalar defined by $\{\varphi_i(s), Z(s)\}$.

It remains for us to show that the equilibrium states belong to a surface of $\mathbb{R}^{m+1}$, which will correspond to the above function $\log a_\theta$.

DEFINITION. *A system is in a state of equilibrium σ if $\forall s \in S_T (\sigma > s)$ $(\sigma > s) \mapsto (s > \sigma)$. That is to say that all subsequent observations of σ are σ.*

THEOREM. *The ensemble of equilibrium states belongs to the upper sheet of the positive cone of* $\mathcal{T}$.

Let us consider: $\{s \in S, s > \sigma\} = \{\sigma\}$;

$\varphi(\{\sigma\})$ is well defined and the co-ordinates of the $s \in \{\sigma\}$ differ only by $Z(s)$. Let us call $a(\{\sigma\}) = \sup_{\{\sigma\}} \{\lambda = Z(ns)/n\}$ and be $\alpha : \mathcal{Q} \to \mathbb{R}$ defined by:

$$\alpha(f) = a\{\varphi(s)\} \qquad \forall s \in \{\sigma\}.$$

Then α is linear convex on $\mathcal{Q}$.

Conclusion

The equilibrium states belong to the graph of α in (φ, Σ); this graph is the manifold of M_c seen above, since Σ has all the properties of $\log a_\theta$. The two models provide isomorphic Boolean lattices.

In the experimental description, the $\theta_i = \partial \log a_\theta / \partial \varphi_i$ correspond to the additive functions (linearity of tangent space) defined on α, and constant on the family of states generated by an equilibrium state.

Then $s_1 + s_2$ is an equilibrium state if s_1 and s_2 are equilibrium states, that is $\theta_i(s_j) = \theta_i(s_k)\ \forall i$.

12. Conclusion

In this study we have attempted to give a correct description of the hypotheses or of the contingent data of the models used in classical mechanics before the introduction of quantum mechanics. On this occasion, the propositions provided by these models and their algebraic structure have been defined, each structure being necessarily embedded in the previous structure, without being a sub-structure. In particular we have passed from a classical Boolean lattice to a lattice of continuous geometry type for the Brownian model, then to a Boolean lattice for the thermodynamic model.

From this viewpoint we can give clearer consideration to the problem of 'hidden variables' of quantum mechanics. The structure of the propositions is a Croc [35b] and no change of variable will affect it; in the same manner no extension, in which the quantum lattice is the quotient lattice by a sub-algebra of hidden variables, is possible, since the quotient of a Boolean algebra bya factor is still Boolean [6a]. Opposite to these models, the extension which consists of imbedding a lattice in another of higher dimension and Boolean, then regarding the 'disturbing' elements as inobservable is still possible but brings with it no physical information. Let us consider for example: the modular orthocomplemented lattice of the propositions associated with a polarization experiment (two-dimensional Hilbert space) and the lattice of the propositions associated with a classical system with two propositions (a, b).

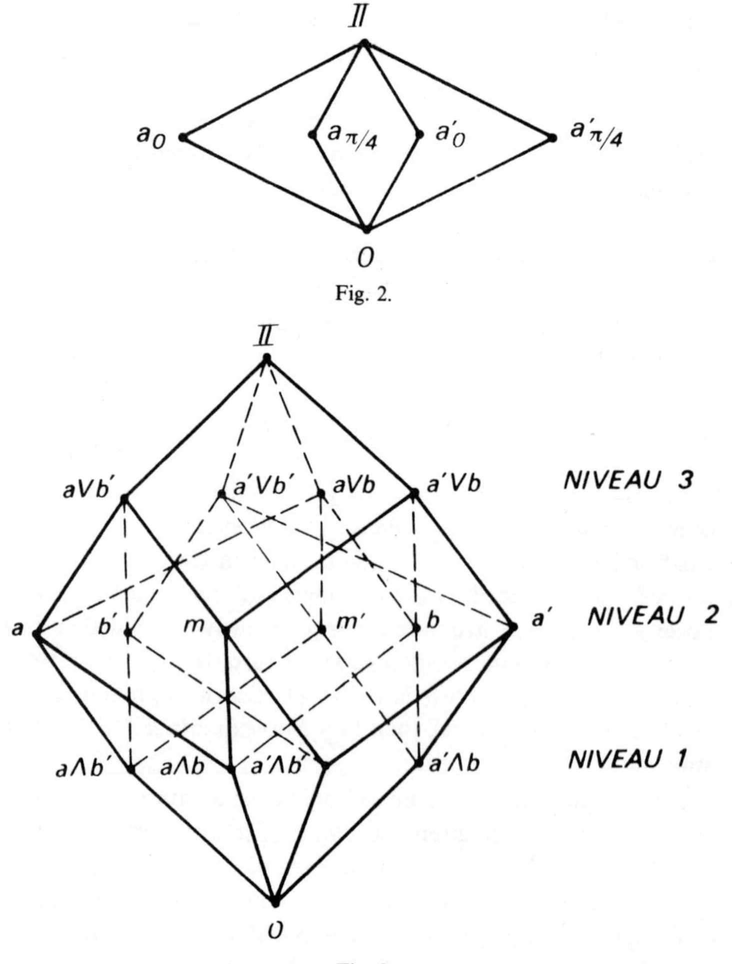

Fig. 2.

Fig. 3.

We can always associate a_0, $a \vee b'$; $a_{\pi/4}$, $a \vee b$; a_0', $a' \vee b'$; $a_{\pi/4}'$, $a' \vee b$ and consider that $(a(\vee b'))'$, $a'(\vee b')$, etc.... and the smaller propositions as unobservable in the quantum model (that is non-preparable). (For example, take an apparatus of which the precision is given by the third level and of which the precisions associated with the lower levels are not realizable). On generalizing this diagram, we can always regard the lattice of a finite dimension Hilbert space as a sub-ensemble of the Boolean lattice of all the parts of the corresponding projective space (rays-space). But this form of extension provides no physical information.

By making use of the extension structure given by the chain of the models described, we can on the other hand envisage an extension of the quantum model having a precise physical meaning. We can also note that this analogue method is at the source of quantum formalism itself.

Institut de Physique Nucléaire,
Université Claude Bernard, Lyon.

NOTE

* Let $Y^i_{n_i}$ be the propositions of $\mathscr{L}_{\mathscr{A}_i}$. To define $Y^i_{n_i} \vee Y^j_{n_j}$ let us consider $\mathscr{L}_B$ generated by union and intersection of all the $Y^i_{n_i}$ for all i. Within $\mathscr{L}_B$, $Y^i_{n_i} \underset{B}{\vee} Y^j_{n_j}$ is well defined. Within $\mathscr{L}_{\mathscr{A}_k}$ the set of upper-bound of $Y^i_{n_i} \underset{B}{\vee} Y^j_{n_j}$ is not empty. Let Z^{ij}_k be the lower bound of these upper-bound of $\mathscr{L}_{\mathscr{A}_k}$. The Z^{ij}_k are ordered by k, where k is associated with the precision of $\mathscr{A}_k$. By putting $\inf_k Z^{ij}_k = Y^i_{n_i} \vee Y^j_{n_j}$ and in the dual manner for $\wedge$, we define the lattice $\mathscr{L}_{\mathscr{A}}$. Orthocomplementation in $\mathscr{L}_A$ is caused by that of the $\mathscr{L}_{\mathscr{A}_i}$.

BIBLIOGRAPHY

1. Anderson, B. O.: *Principles of Relativity Physics* (Chapter 1), Academic Press, New York, 1967.
2. Araki, H. and Yanase, M. M.: *Phys. Rev.* **120**, 622 (1960).
3. Arnesov, W. B.: *Acta Math.* **118**, 95 (1967).
4. Arnold, V. I. and Avez, A.: *Problèmes ergodiques de la mécanique classique*, Gauthier-Villars, Paris, 1967.
5. Bachelard, G.: *L'activité de la physique rationaliste* (chap. 2, parag. 7), Presses Univ. de France, Paris, 1966.

6a. Birkhoff, G.:
   (a) *Lattice Theory*, Am. Math. Soc., fasc. XXV, 1967, p. 282.
   (b) *Bull. Am. Math. Soc.* **50**, 764 (1944).
   (c) *Lattice Theory*, Am. Math. Soc., fasc. XXV, 1967, p. 260.
   (d) *Dynamical Systems*, Am. Math. Soc., Publ. No. IX, 1927.

6b. Boccara, N.: *Les principes de la thermodynamique classique*, Presses Univ. de France, 1968.

7. Cameron, R. H. and Martin, W. T.: *Ann. Math.* **48** 385 (1947).
8. Chandrasekhar, S.: *Rev. Mod. Phys.* **15**, 16 (1943).
9. Cherry, T. M.: *Proc. Cambr. Phil. Soc.* **22**, 287 (1925).
10. Collins, R. E.: *Phys. Rev.* **183**, 1081 (1969).
11. Doob, J. L.: *Stochastic Processes*, Wiley and Sons, New York, 1964, p. 74.
12. Dunford, N. and Schwartz, J. T.: *Linear Operators*, Vol. 1, Interscience Publ., New York, 1964, p. 62.
13. Einstein, A., Rosen, N., and Podolsky, B.: *Phys. Rev.* **47**, 777 (1935).
14. Fischer, R. (Sir): *Le plan d'expérience*, Editions du C.N.R.S., Paris, 1961, No. 110.
15. Gréa, J.: Thèse (1974), Lyon, p. 44.
16. Griffiths, R. G.: *J. Math. Phys.* **6**, 1447 (1965).
17. Hagiara, Y.: *Celestial Mechanics*, Vol. 1, MIT Press, Cambridge, 1970, p. 305.
18. Halmos, P. R.:
    (a) *Lectures on Ergodic Theory*, Chelsea Publ., New York, 1958.
    (b) *Measure Theory*, Von Nostrand Co., Amsterdam, 1965.
    (c) *Bull. Am. Math. Soc.*, **55**, No. 11 (1949).
19. Hemmer, P. Chr.: *Dynamic and Stochastic Types of Motion in the Linear Chain*, Ph.D. Thesis, Univ. Trondheim, Norvège, 1959.
20. Hope, E.:
    (a) *Ergoden Theorie*, Springer-Verlag, Berlin, 1937.
    (b) *J. Math. Phys.* **13**, 51 (1934).
21. Khinchin, A. I.:
    (a) *Mathematical Foundations of Statistical Mechanics*, Dover Publ., New York, 1949.
    (b) *Idem*, réf. (21 a), p. 52.
    (c) *Mathematical Foundations of Quantum Statistics*, Dover Publ., New York, 1960.
22. Koopman, B. O. and Von Neumann, J.: *Proc. Nat. Acad. Sci.* **18**, 255 (1932).
23. Koopman, B. O.: *Proc. Nat. Acad. Sci.* **17**, 315 (1931).
    Trans. Am. Math. **39**, 399 (1936).
24. Kubo, R.: *Thermodynamics*, North-Holland Publ., Amsterdam, 1968, p. 136.
25. MacLaren, M. D.: *Notes on Axioms for Quantum Mechanics*, Rapport ANL-7065.
26. Landau, L. and Lifchitz, Y.: *Physique statistique*, Editions MIR, Moscou, 1967, parag. 113–114.
27. Lewis, R. M.: *Arch. Rat. Mec. Anal.* **5**, 355 (1960).
28. Mackey, G. W.: *Mathematical Foundations of Quantum Mechanics*, Benjamin Inc., Londres, 1963.
29. Mielnik, B.: *Comm. Math. Phys.* **15**, 1 (1969).
30. Misra, B.: *Nuovo Cimento* **47**, 841 (1967).
31. Nikodym, O. M.: *The Mathematical Apparatus for Quantum Mechanics*, Springer-Verlag, Berlin, 1966.
32. Von Neumann, J. V.:
    (a) *Ann. Math.* **102**, 110 (1929).
    (b) *Proc. Nat. Acad. Sci.* **18**, 70 (1932).
33. Onofri, E. and Pauri, M.: *J. Math. Phys.* **14**, 1106 (1973).
34. Penrose, O.: *Foundations of Statistical Mechanics*, Pergamon Press, New York, 1970.
35. Piron, C.:
    (a) *Règles de supersélection continue*, Inst. Phys. Théor., Genève.
    (b) *Helv. Phys. Acta* **37**, 439 (1964).
    (c) *Helv. Phys. Acta* **104**.
36. Piron, C. and Jauch: *Helv. Phys. Acta* **36** 887 (1963).

37a. Ruelle, D.: *Statistical Mechanics*, Benjamin Inc., Londres, 1969.
37b. Sikorski, R.: Vol. II, *Boolean Algebras*, Springer-Verlag, Berlin, 1964, parag. 38.
38. Siegel, C. L. and Moser, J. K.: *Lectures on Celestial Mechanics*, Springer-Verlag, Berlin, 1971.
39. Szasz, G.:
(a) *Théorie des treillis*, chap. X, parag. 64;
(b) *Théorie des treillis*, chap. II, parag. 13;
(c) *Théorie des Treillis*, pp. 129–130, Théor. 62;
Dunod, Paris, 1971.
40. Wang, M. C. and Uhlenbeck, G. E.: *Rev. Mod. Phys.* **17**, 323 (1945).
41. Wiener, N.: *Non Linear Problems in Random Theory*, MIT Press, Cambridge, 1958.
42. Wiener, N. and Siegel, A.: *Phys. Rev.* **91**, 1551 (1953).
43. Zierler, N.: *Pac. J. Math.* **II**, 1152 (1961).

C. PIRON

ON THE FOUNDATIONS OF QUANTUM PHYSICS

The interpretation of quantum theory has always been a source of difficulties, especially with regard to the theory of measurement. We do not intend to enter here into the details of the polemic which has surrounded this problem. The article by London and Bauer [1] is an excellent introduction to the subject, and Wigner's masterly exposition [2] which opened the summer-school of Varenna in 1970, clearly brings into evidence different points of view in the controversy.

In our opinion, this polemic shows that there seems to be an irreducible opposition between the language of classical physics and the language of quantum physics [3]. This opposition between the two languages is particularly irritating because in quantum physics one must make an appeal to classical concepts to describe the measurements. Also, if one considers a pure quantum description of the evolution of a macroscopic system, one is sometimes led to paradoxical results, contradicting our picture of reality. 'Schrödinger's cat' [4] is a well-known example of such a paradox.

In general one tries to avoid these difficulties by either one of the following two proposals:

(i) one renounces the description of individual systems, and concedes reality only to the statistical results expressed in terms of probability with respect to an ensemble. In our opinion, this is the case for the theory proposed by Ludwig [5] and also the one proposed by Everett-Wheeler-DeWitt [6].

(ii) one tries to return to a 'classical' description, at the cost of definitively renouncing any attempt at complete description. This is the case for the hidden variable theories.

Both of these approaches are incompatible with the 'program of realism' which one may formulate as follows:

> to give a complete description of each individual system as it is in all its complexity.

M. Flato et al. (eds.), Quantum Mechanics, Determinism, Causality, and Particles, 105–116. *All Rights Reserved. Copyright © 1976 by D. Reidel Publishing Company, Dordrecht-Holland.*

The difficulties with and the strangeness of the quantum language may lead one to believe that it is impossible to realize such a program even approximately. However, it is perfectly possible to describe completely a system in terms of its properties, i.e. its elements of reality (in the sense of Einstein), and this without making use of the notion of probability; one can comprise in the same formalism the classical theory and the quantum theory, and thus clarify the role of the linear structure. We shall outline such a description in the following paragraphs.

1. The Propositional System

The mathematical category which permits such a description is that of CROCs.

DEFINITION. A CROC is a set $\mathscr{L}$ endowed with the two structures:

(1) A partial ordering relation denoted $a<b$, according to which $\mathscr{L}$ is a complete lattice: for all subsets of elements $a_a \in \mathscr{L}$ there exists a least upper bound $\bigvee_a a_a \in \mathscr{L}$, and a greatest lower bound $\bigwedge_a a_a \in \mathscr{L}$. In particular, there exists a minimal element 0 and a maximal element I.

(2) A mapping from $\mathscr{L}$ onto $\mathscr{L}$, called an orthocomplementation, which

(i) for each $a \in \mathscr{L}$ associates a complement denoted a' such that

$$a \wedge a' = 0 \quad \text{and} \quad a \vee a' = I,$$

(ii) is an involution:

$$a<b \Rightarrow b'<a', \quad a''=a,$$

(iii) makes $\mathscr{L}$ weakly modular

$$a<b \Rightarrow a \vee (a' \wedge b) = b.$$

DEFINITION. A morphism in the category of CROCs is a mapping μ from a CROC $\mathscr{L}_1$, into a CROC $\mathscr{L}_2$ such that

(i) $\mu(\bigvee_a a_a) = \bigvee_a \mu a_a$,

(ii) $a<b' \Rightarrow \mu a < (\mu b)'$.

Before constructing such an object in physics let us discuss the notion of *question* and *proposition*.

DEFINITION. One calls a question every experiment leading to an alternative of which the terms are 'yes' and 'no'.

There exists a trivial question which we shall denote as I, and which consists in measuring nothing and stating the answer 'yes' each time.

When the physical system has been prepared in such a way that the physicist may affirm that in the event of an experiment the result 'yes' is certain, we shall say that the question is *certain*, or that the question is *true*.

DEFINITION. One says that a question α is *stronger* than a question β and one denotes this with $\alpha < \beta$, if whenever α is true, one can affirm that β is true.

This relation is transitive, therefore it is a quasi-order and it induces an equivalence relation:

$$\alpha \sim \beta \Rightarrow \alpha < \beta \quad \text{and} \quad \beta < \alpha.$$

DEFINITION. One calls a *proposition* and denotes by a, the equivalence class containing the question α.

A proposition a is denoted as *true* for a given physical system if and only if one of the questions $\alpha \in a$ is true. This corresponds to a property, i.e. an element of reality, for the physical system.

To construct the CROC $\mathscr{L}$ associated to a given kind of physical system we proceed in the following way: first we define from the set $\mathscr{L}$ of propositions a partial ordering relation

$$a < b \Rightarrow \alpha < \beta \quad \text{for} \quad \alpha \in a, \ \beta \in b\,;$$

that is to say 'a true' $\Rightarrow$ 'b true'.

THEOREM. $\mathscr{L}$ is a complete lattice for this partial order.

Proof. Let a family of propositions $b_i \in \mathscr{L}$ be given, and choose a representative β_i for each b_i. Then we can define a new question $\prod_i \beta_i$ by the following prescription: One chooses one of the β_i in the family arbitrarily, one performs the corresponding experiment, and one attributes to $\prod_i \beta_i$ the answer thus obtained. It is easy to verify that $\prod_i \beta_i$ is true if and only if all the β_i are true. The equivalence class containing $\prod_i \beta_i$ defines a new proposition $\bigwedge_i b_i$ which is the greatest lower bound for the b_i.

The existence of a least upper bound follows immediately from the existence of a greatest lower bound for the family

$$m = \{x \mid x \in \mathscr{L} \quad \text{and} \quad b_i < x, \forall_i\}$$

by

$$\bigvee_i b_i = \bigwedge_{x \in m} x.$$

Secondly, we have to define an orthocomplementation on the lattice of propositions $\mathscr{L}$. For this, we need the notion of *compatible complement.*

DEFINITION. Two propositions a and b are *compatible complements* for each others if

(i) they are complements for each other's

$$a \wedge b = 0 \quad \text{and} \quad a \vee b = I$$

(ii) if there exists a question $\alpha \in a$ such that the inverse question $\alpha^{\sim}$, obtained by exchanging the terms of the alternative, is contained in b.

Furthermore, we must impose the following two axioms:

AXIOM C. For each proposition a of the lattice $\mathscr{L}$ there exists at least one compatible complement denoted a'.

AXIOM P. If $a < b$ are propositions of the lattice $\mathscr{L}$ and if a' and b' are compatible complements for a and b respectively, then the sub-lattice generated by $\{a, a', b, b'\}$ is distributive.

Let us recall that a lattice $\mathscr{L}$ is distributive if and only if for each triplet $a, b, c \in \mathscr{L}$ we have

$$a \vee (b \wedge c) = (a \vee b) \wedge (a \vee c)$$

In classical physics the lattice of propositions corresponds to the lattice of subsets of the phase-space, and it is a distributive lattice. This characteristic property of the classical systems permits one to understand the meaning of axiom P.

THEOREM. A lattice $\mathscr{L}$ satisfying the axioms C and P is a CROC; and conversely, if one interprets the orthocomplement in a CROC as a compatible complement, then the axioms C and P are satisfied.

Proof. If $\mathscr{L}$ satisfies the axioms C and P then we have:

$$a<b \Rightarrow b'<a'$$

since by the distributivity:

$$a'=a' \vee (b' \wedge b)=(a' \vee b') \wedge (a' \vee b)=a' \vee b'$$

Thus, the compatible complement is unique and the mapping $a \mapsto a'$ is an orthocomplementation. Finally, $\mathscr{L}$ is weakly modular:

$$a<b \Rightarrow a \vee (a' \wedge b)=b$$

Conversely, in a CROC, C is trivial and for P we remark that for $a<b$ one has to show that the sub-lattice generated by $\{a, a', b, b'\}$ is distributive. This amounts to the verification that the two relations

$$a \vee (a' \wedge b)=b \quad \text{and} \quad b' \vee (b \wedge a')=a'$$

are necessary and sufficient for

$$\{0, a, a' \wedge b, b, b', a \vee b', a', I\}$$

to be a distributive sub-lattice.

The axioms C and P permit us to define the following important concept.

DEFINITION. Two propositions a and b in a CROC are said to be compatible ($a \leftrightarrow b$) if and only if the sublattice generated by $\{a, a', b, b'\}$ is distributive.

The set $\mathscr{P}(H)$ of projectors P_a of a Hilbert space is a CROC under the partial ordering

$$P_a < P_b \Leftrightarrow P_a P_b = P_a$$

and the orthocomplementation

$$P_a \mapsto P_{a'} = I - P_a.$$

It is also easy to show that two projectors are compatible if and only if they commute.

The state of a particular physical system (prepared in a given way) is defined by the complete set of its elements of reality, i.e. the sub-set $S \subset \mathscr{L}$ of all the propositions which are true for the system under consideration.

In classical theory as in quantum theory, such a set is identical with the set of propositions which majorizes an atom i.e. a proposition $p \neq 0$ for which

$$0 < x < p \Rightarrow x = 0 \quad \text{or} \quad x = p.$$

Thus the states of a system are in one-to-one correspondence with the atoms of the CROC. On the other hand, by definition, it must be possible, for each proposition $a \neq 0$, to prepare the system in such a way that 'a is true'. This remark justifies the following mathematical axiom.

AXIOM A_1. The CROC of propositions is atomic: for every proposition $a \neq 0$ there exists an atom $p < a$.

For completeness let us also mention the last axiom which characterizes the CROC of propositions.

AXIOM A_2. If $a \neq 0$ and p is an atom, and if $a' \wedge p = 0$ then $(p \vee a') \wedge a$ is an atom.

DEFINITION. One calls a CROC satisfying the axioms A_1 and A_2 a *propositional system.*

The CROC of sub-sets of phase-space and the CROC of projectors of Hilbert-space are examples of propositional systems.

An observable is a correspondence between the propositions defined by the scale of the measuring apparatus and certain propositions of the system. Such a correspondence permits one to affirm a property of the system from the knowledge of a property of the apparatus. This justifies the following mathematical definition:

DEFINITION. A morphism

$$\mu : \mathcal{B} \to \mathcal{L}$$

from a distributive CROC $\mathcal{B}$ into the propositional system $\mathcal{L}$ is called an observable. Usually, one also imposes the additional condition $\mu I = I$.

Such a definition recovers the usual notion of observable. In fact, one can show, as we have done elsewhere [7]:

(i) that in the classical case, each observable can be defined by the inverse image of a function on the phase-space; and conversely, each such function defines an observable.

(ii) that in the usual quantum case, to each observable corresponds a self-adjoint operator whose spectral family (of projectors) defines the morphism; and conversely, to each self-adjoint operator corresponds a spectral family which defines an observable.

2. General formalism and superselection rules

The *center* of a propositional system is by definition the set of propositions compatible with all the others. Let us recall some results [8]. The center is by itself a distributive proportional system, and therefore, it is isomorphic to the CROC of the sub-set of a set. In the purely classical case, the center is the whole propositional system and so one justifies the use of the sub-sets of the phase-space in classical physics. In the purely quantum case (for example in the usual wave-mechanics) the center reduces to the propositions 0 and I. Then the propositional system is essentially the CROC of projectors of a Hilbert space.

In the general case, where the center is neither $\{0, I\}$ nor the whole propositional system, one says that the system possesses *superselection rules*, and in this case it is a product (in the sense of the categories) of irreducible propositional systems, i.e. of the purely quantum kind. In the following, we will consider a particular case which is sufficiently general to describe both classical systems and usual quantum systems.

Let us consider a family of separable complex Hilbert spaces H_a where $\alpha \in \Omega$, a set of indices, and define the propositional system $\mathscr{L}$ as the product $\bigvee_a \mathscr{P}(H_a)$. Then a proposition is represented by a family $\{P_a\}$ of projectors. The ordering relation

$$\{P_a\} < \{Q_a\}$$

is given by

$$P_a Q_a = P_a, \quad \forall \alpha \in \Omega$$

and the orthocomplementation by the mapping

$$\{P_a\} \mapsto \{I - P_a\}.$$

Two propositions $\{P_a\}$ and $\{Q_a\}$ are compatible if and only if

$$[P_a, Q_a]=0, \quad \forall\alpha\in\Omega.$$

The state of $\bigvee_a \mathscr{P}(H_a)$ is described by giving a ray in one of the Hilbert spaces H_a, or equivalently, by giving a family of projectors all of which are null except for one which is of rank one. Each observable is defined by giving a family $\{A_a\}$ of self-adjoint operators, and each symmetry (i.e. automorphism) is induced by a permutation f of the elements of Ω and a family of unitary (or anti-unitary) operators

$$U_a : H_a \to H_{fa}.$$

Finally, to be complete, let us consider the problem of the description of the dynamics. We assume that a reversible evolution is induced by a representation of the one-parameter group of translations in the symmetries of the propositional system. Therefore, we associate to each interval of time τ a permutation

$$\alpha \mapsto \alpha_\tau$$

and a family of unitary operators

$$U_a(\tau): H_a \to H_{a_\tau},$$

together satisfying the relations

$$(\alpha_{\tau_1})_{\tau_2} = \alpha_{\tau_1+\tau_2}$$

and

$$U_{a_{\tau_1}}(\tau_2)\, U_a(\tau_1) = U_a(\tau_1+\tau_2).$$

If we also assume some conditions of continuity and differentiability we can deduce the following equations

$$\partial_\tau \alpha_\tau = x(\alpha_\tau)$$

and

$$i\partial_\tau \psi_{a_\tau} = \mathscr{H}_{a_\tau} \psi_{a_\tau}$$

i.e.: a Schrödinger equation coupled with a system of differential equations defined by a vector-field.

3. The Problem of Measurement and of Irreversible Evolution

If we want to describe the perturbation undergone by the physical system during a measuring process it is quite natural to consider the system and the measuring apparatus together, as one system, and we will refer to this as the *composite system*. We may expect to be able to describe such a composite system in the formalism already outlined; because, if the experience shows that a certain quantity defined for the measuring apparatus (for example the position of its pointer) has a well-defined value under all circumstances, and varies continuously with time, it is sufficient for avoiding any paradox to consider this quantity as one of the variables defining Ω, i.e. as a superselection rule. There is no difficulty of principle in doing this in our formalism, although it is rigorously impossible in the usual one, since in this case the superselection rules if they exist are discrete and correspond to constants of motion.

Thus having the possibility to describe the state of the composite system at each instant, one must for the sake of completeness also describe its dynamics. It is easy to see that it is not possible to explain the particular evolution of such a system by the Schrödinger type equations of the last paragraph. In fact, according to these equations the evolution of the vector ψ_{a_τ} depends explicitly on the values of $\alpha_\tau \in \Omega$; however, the evolution of the α_τ themselves (in this case the position α of the pointer) do not depend on the vector given in H_{a_τ}. Thus, since by definition a measurement must establish a correlation between some vectors of one space H_{a_0} and different values of α, we arrive at the above conclusion. Physically, this conclusion is not surprising since a measuring process is irreversible. If we are content to explain the evolution of the system by irreversible equations, it is not difficult to find such, but rather to choose one of the many examples known in physics.

Finally, to show the power of our formalism, we will treat an example of an irreversible process. It concerns a singlet state which disintegrates into two identical particles of mass m and spin $\frac{1}{2}$. In our model we do not describe the motion of the center of mass, and we assume that the observables of the relative momentum and the relative position are superselection rules. Thus only the two spins are quantified. The corresponding propositional system is then defined by a family $\{H_{\mathbf{p},\mathbf{q},t}\}$ of Hilbert spaces, labelled by the variables relative momentum $\mathbf{p}$, relative position $\mathbf{q}$ and time t; each space $H_{\mathbf{p},\mathbf{q},t}$ being canonically isomorphic to the four dimensional complex Hilbert space $\mathbb{C}^2 \otimes \mathbb{C}^2$. The initial state just after the disintegration is given by

$$\mathbf{p}_0 = \mathbf{p}$$

$$\mathbf{q}_0 = \mathbf{0}$$

$$t_0 = 0$$

$$\psi_{\mathbf{p}_0, \mathbf{q}_0, t_0} = \sqrt{\tfrac{1}{2}}(|+-\rangle - |-+\rangle) = \sqrt{\tfrac{1}{2}} \begin{bmatrix} 0 \\ 1 \\ -1 \\ 0 \end{bmatrix}$$

The evolution of such a system in an external magnetic field parallel to the z-axis is, if it is reversible, given by

(i) the classical Hamiltonian $\mathscr{H}(\mathbf{p}; \mathbf{q}, t) = \mathbf{p}^2/m$ giving the equations

$$\dot{\mathbf{p}}_t = 0 \quad \text{and} \quad \dot{\mathbf{q}}_t = \frac{2}{m} \mathbf{p}_t,$$

(ii) and the quantal Hamiltonian

$$\mathscr{H}_{\mathbf{p},\mathbf{q},t} = -\gamma\left(B\left(-\frac{\mathbf{q}}{2}\right) \sigma_z \otimes I + B\left(+\frac{\mathbf{q}}{2}\right) I \otimes \sigma_z \right)$$

defining the Schrödinger equation

$$i\partial_t \psi_{\mathbf{p},\mathbf{q},t} = \mathscr{H}_{\mathbf{p},\mathbf{q},t} \psi_{\mathbf{p},\mathbf{q},t}$$

Now, the initial state $\psi_{\mathbf{p}_0,\mathbf{q}_0,t_0}$ is an eigen-vector for $\mathscr{H}_{\mathbf{p},\mathbf{q},t}$ if and only if

$$B\left(-\frac{\mathbf{q}}{2}\right) = B\left(\frac{\mathbf{q}}{2}\right)$$

This is only the case for $\mathbf{q} = 0$, that is at the instant $t = 0$. Thus this initial state is not stationary and the system will in fact show relaxation effects. Such effects can be described with the help of the von Neumann density matrix. One is thus led to write some equations for the elements of the density matrix, which in the representation diagonalizing the Hamiltonian take the following form [10]

$$i\partial_t \rho_{ij}(t) = [\mathscr{H}_{\mathbf{p},\mathbf{q},t}, \rho(t)]_{ij} - i\frac{1}{T}(1-\delta_{ij})\, \rho_{ij}(t)$$

If the relaxation time T is much greater than the considered time interval, we can neglect the relaxation terms. However, in the opposite case the system evolves towards the final matrix:

$$\rho_\infty = \tfrac{1}{2}\begin{bmatrix} 0 & 0 & 0 & 0 \\ 0 & 1 & 0 & 0 \\ 0 & 0 & 1 & 0 \\ 0 & 0 & 0 & 0 \end{bmatrix}$$

corresponding to a mixture of the states $|+-\rangle$ and $|-+\rangle$.

In conclusion, the interpretation of the quantal phenomena requires us neither to give up the classical realistic point of view nor to invent new 'ad hoc' languages. In fact, the particularity of classical physics comes from the purely accidental possibility, in all reasoning to identify the result of an eventual action on the system with an observed (or actual) fact.

This being said, the irreversible dynamics which characterize the real evolution of a system give rise to the same difficulties in quantum theory as occurred in classical theory. And the example that we have presented manifestly shows that the hidden interaction (or, if one wants, the hidden variables) instead of explaining the quantal phenomena are the cause of their usual disappearance on the macroscopic level. This confirms the view that it is the accuracy of our measurements and not the defects of our apparatus which permits us to exhibit quantal effects.

Université de Genève

BIBLIOGRAPHY

[1] London, F. and Bauer, E., *La théorie de l'observation en mécanique quantique*. Hermann, Paris, 1939.

[2] Wigner, E., *The Subject of Our Discussions*, Proceedings of the International School of Physics "Enrico Fermi", Il Corso; *Foundations of Quantum Mechanics*: Academic Press Inc., New York, 1971.

[3] Bohn, D., 'Quantum Theory as an Indication of a New Order in Physics', *loc. cit.* [2].

[4] Schrödinger, E., *Naturwiss.* **48** (1935), 52.

[5] Ludwig, G., *Deutung des Begriffs physikalische Theorie und axiomatische Grundlegung der Hilbertraumstruktur der Quantenmechanik durch Hauptsätze des Messens*. Springer-Verlag Berlin, Heidelberg, 1970.

[6] Everett III, H., *Rev. Mod. Phys.* **29** (1957), 454; Wheeler, J. A., *Rev. Mod. Phys.* **29** (1957), 463; DeWitt, B. S., 'The Many Universes Interpretation of Quantum Mechanics', *loc. cit.* [2].
[7] Piron, C., 'Observables in General Quantum Theory', *loc. cit.* [2].
[8] Piron, C., *Helv. Phys. Acta* **37** (1964), 439; *Found. Phys.* **2** (1972), 287.
[9] For a model of the process of disintegration, see:
Piron, C., *Helv. Phys. Acta* **42** (1969), 330.
[10] See for example p. 11 in:
Pantell, R. H. and Puthoff, H. E., *Fundamentals of Quantum Electronics*, John Wiley and Sons Inc. New York, 1969.

BOGDAN MIELNIK

QUANTUM LOGIC: IS IT NECESSARILY ORTHOCOMPLEMENTED?

1. Introduction

One of the intriguing problems of the present day theory is the lack of similarity between general relativity and quantum mechanics. General relativity is a product of a long evolution line of classical theories leading toward structural flexibility. The most characteristic steps of that evolution were: (1) the discovery of space-time geometry (stage of Minkowski space), (2) the generalization of the geometry (introduction of the pseudo-Riemannian manifolds), and (3) the discovery that geometry depends on matter. In spite of its classical character general relativity is an example of an evolved theory: its fundamental structure is not given a priori (apart from generalities concerning the category) but is conditioned by physics.

The development which led to quantum theories was not similar to that. Here, there was only one decisive step: the abandoning of causal schemes and the transition to the probabilistic wave mechanics. Subsequent progress consisted in improving the symbolic language of states and observables sufficiently to include probabilistic information of increasing complexity. In spite of its rapid development the quantum theory did not undergo any further intrinsic changes of fundamental character and has not achieved a structural flexibility analogous to that of general relativity. Similarly, as in the twenties, present day quantum mechanics represents the variety of possible physical situations (pure states) by the same standard mathematical structure which is the unit sphere in a separable Hilbert space. Unlike the Riemannian manifolds the quantum mechanical unit spheres do not differ one from another: they are all isomorphic. The worlds of the present-day quantum mechanics thus present a picture of structural monotony: they are all 'painted' on that same standard ideally symmetric surface. The formalism of the quantum theory of infinite systems and quantum field theory is not very different from that. In spite of several mathematical refinements

M. Flato et al. (eds.), Quantum Mechanics, Determinism, Causality, and Particles, 117–135. *All Rights Reserved. Copyright © 1976 by D. Reidel Publishing Company, Dordrecht-Holland.*

(the introduction of C*-algebras, the Gelfand-Segal construction) the basic structural framework of the theory is conserved at the cost of quantitative multiplication: when meeting a new level of physical reality the quantum theory responds by simply producing infinite tensor products of its basic structure. The resulting development is more similar to an expansion than to an intrinsic evolution: one just submits other branches of physical theory to the standard language of states and observables which has almost become the only acceptable way of thinking in quantum theories. A still unfinished stage of that expansion process is the programme of gravity quantization. A somewhat disquieting question arises, however: is the structure of the present day quantum theory indeed general enough to assure that further progress may be achieved just by continuing the techniques of operators in Hilbert spaces? Or, perhaps, the situation is different. It may be that present day quantum theory still represents a relatively primitive stage of development and lacks some essential evolutionary steps leading towards structural flexibility [2]. If this were so, further development would involve a programme opposite to the 'quantization of gravity': instead of modifying general relativity to fit quantum mechanics one should rather modify quantum mechanics to fit general relativity. The way toward flexible quantum structures was recently opened in the convex set theoretical approach to quantum mechanics [1, 3, 5, 6]. On the other hand, there exist conservative arguments supporting the necessity of the present day form of quantum theory, which are found in the axiomatics of quantum logic [4, 7]. This is why the axioms of quantum logic should be critically reexamined.

2. Lattice of Macroscopic Measurements

According to a generally accepted philosophy the 'quantum logic' is the set of all 'questions' which may be put to micro-object. By a *question* (also: *proposition, yes-no measurement*) one usually understands any physical arrangement which, when interacting with a microobject, may or may not produce a certain macroscopic effect interpreted as the answer 'yes'. Though the 'question' may be put to any single micro-object, the answer becomes conclusive only if obtained for a great number of its independent replies. This leads to an abstract scheme where 'questions' idealize the macroscopic devices used to test the statistical ensembles of microsystems. Let now Q denote the set of all 'questions' for certain definite physical objects. For completeness it will be assumed that Q contains two trivial questions: 'I' to which the answer is always 'yes' and '0' to which the answer 'yes' is never given. The existence of statistical ensembles as the counterpart of Q allows us to introduce a certain structure in Q which is the most recognized element of geometry in quantum theory.

DEFINITIONS. Given an ensemble x and a question $a \in Q$, one says that the answer 'yes' to the question a is certain for the individuals of x if 'yes' is obtained for the average fraction 1 of the individuals of that ensemble. The ensembles x for which the answer 'yes' to the question a is certain will be told to form the 'certainly yes domain' of a. Given two questions $a, b \in Q$, we say that a is more restrictive than $b (a \leqslant b)$ if the certainty of the answer 'yes' to the question a implies the certainty of 'yes' to the question b. Thus, $a \leqslant b$ if the 'certainly yes' domain of a is contained in the 'certainly yes' domain of b.

The relation $\leqslant$ is reflexive and transitive. The further properties of $\leqslant$ are associated with the 'logical' interpretation. According to that interpretation the questions $a \in Q$ represent the elements of an abstract 'logic' which reflects the nature of the microsystems: the relation $\leqslant$ is the implication of the logic. Since in any logical system the pair of implications $a \Rightarrow b$ and $b \Rightarrow a$ means that 'a is equivalent to b', one generally assumes that a similar property should hold in Q.

AXIOM I. Two questions $a, b \in Q$ with identical 'certainly yes' domains are physically equivalent (i.e., cannot be distinguished by observing how they select any statistical ensemble). Formally:

$$a \leqslant b \quad \text{and} \quad b \leqslant a \Rightarrow a = b. \tag{2.1}$$

In consequence, the relation $\leqslant$ introduces a partial order in Q with upper and the lower bounds I and 0. Because of common experience of classical and quantum phenomenology one also assumes that the partial ordering $\leqslant$ makes Q a lattice.

AXIOM II. For every pair $a, b \in Q$ the partial order $\leqslant$ determines the unique lowest upper bound $a \vee b \in Q$ called the *union* of a and b. Similarly, for every $a, b \in Q$ there exists in Q the greatest lower bound $a \wedge b$ called the *intersection* of a and b.

The physical interpretation given to the union $a \vee b$ is that of an experimental arrangement which yields the answer 'yes' with certainty for those systems for which either a or b give certainly the answer 'yes'. Similarly, $a \wedge b$ is interpreted as an arrangement which yields the answer 'yes' with certainty if both a and b yield the answer 'yes' certainly. The 'logical' interpretation given to the operations $\wedge$ and $\vee$ is that of conjunction and alternative. Since Q is a 'logic', and the logical systems admit negation, one generally assumes the following axiom about an orthocomplemented nature of Q:

AXIOM III. There exists in Q a mapping $a \to a'$ which to every $a \in Q$ assigns precisely one $a' \in Q$ called a *negative* of a, such that:

$$a \leqslant b \Rightarrow b' \leqslant a' \tag{2.2}$$

$$a \wedge a' = 0; \ a \vee a' = \mathrm{I} \tag{2.3}$$

$$(a \vee b)' = a' \wedge b'; \ (a \wedge b)' = a' \vee b' \tag{2.4}$$

$$(a')' = a. \tag{2.5}$$

The physical interpretation given to the mapping $a \to a'$ is consistent with the general idea that the question is an arbitrary macroscopic arrangement producing certain macroscopic alternative effects of which one is called 'yes' and the other is 'no'. Now, if a is an arrangement of that kind, the a' is interpreted as essentially the same arrangement with an opposite convention determining what is 'yes' and what is 'no'.

The set of questions Q with the lattice operations $\wedge$, $\vee$ and orthocomplementation $a \to a'$ is sometimes considered the fundamental structure of quantum theory reflecting the nature of the corresponding physical objects. In case of classical objects the questions $a \in Q$ correspond to the subsets of a classical phase space. The symbols $\leqslant$, $\vee$ and $\wedge$ then have the sense of theoretical inclusion, union, and product respectively, while the mapping $a \to a'$ is the operation of taking the set theoretical complement. In that case the logic Q, apart of properties listed in Axioms I, II, III fulfills the distributive law:

$$a \wedge (b \vee c) = (a \wedge b) \vee (a \wedge c). \tag{2.6}$$

One thus infers that the distributive property of the logic Q is a manifestation of the classical nature of the corresponding physical objects. A different case of a 'logic' is obtained by analyzing the structure of orthodox quantum mechanics. Here the 'questions' are the self adjoint operators with two-point spectrum $\{0, 1\}$ in a certain Hilbert space $\mathscr{H}$ (orthogonal projectors). Hence, there is one-to-one correspondence between the elements $a \in Q$ and closed vector subspaces of $\mathscr{H}$. The closed vector subspaces in $\mathscr{H}$ form an orthocomplemented lattice which is not distributive. Hence one infers that in the micro-world classical logic is no longer valid, but a new type of logic becomes relevant in which the alternative is not distributive with respect to the conjunction. One consistently interprets the non-distributive property of Q as the main sign of a non-classical character of the corresponding objects.

3. Motivation of Hilbert Space Formalism

For a certain time the 'quantum logic' Q was considered to be the fundamental structure of quantum theory and has been studied to provide information concerning the most general form possible of quantum mechanics. According to a general belief, the answer should be obtained in the framework of some universal axioms which should reflect the nature of the macroscopic 'yes-no' effects and thus should be valid for any quantum system. The problem of an axiom which would replace the distributive law of classical logic was studied by Piron [7]. He postulated the following property of weak modularity as the one which holds for both orthodox classical and orthodox quantum systems:

$$a \leqslant b \Rightarrow a \vee (a' \wedge b) = b. \tag{3.1}$$

Piron's axiom has no immediate physical interpretation. However, it has been additionally clarified by Pool [8] who has shown that (3.1) is a necessary condition which allows us to associate uniquely the elements $a \in Q$ with some idempotent operations upon the statistical ensembles which represent the selection processes performed by the corresponding measuring devices. In Piron's scheme the weak modularity has been completed by axioms of atomicity and covering [7]. On that basis an important theorem was proved [7]: every irreducible 'quantum logic' must be isomorphic to the lattice of closed vector subspaces in a Hilbert space over one of three basic number fields (real numbers, complex numbers or quaternions). Every reducible quantum logic is a simple product of Hilbert space lattices and thus, corresponds to the orthodox theory with superselection rules.

The above results have a certain unexpected feature. They provide a good structural description of the existing theory. However, they seem to exclude the possibility of generalizations: we return here to the well known scheme of states and observables with the Hilbert space at the bottom [7, 8]. Moreover, the scheme of Piron and Pool is so compact that it is difficult to see in which point it could be relaxed without denying something very fundamental. This is sometimes taken as an argument against the possibility of further generalizations of the present day quantum scheme. However, the conclusion from the lattice theoretical results [7, 8] might be just the opposite. After all, most of the essential progress in physics has been achieved by denying something apparently obvious. Thus, general relativity denied the axioms of Euclid. Present day quantum mechanics has denied the even more obvious distributive law. There is no reason to think that this process is ended. The theorems of Piron and Pool exhibit a conservative quality of quantum logical axioms: it may thus be, that these axioms are the next 'obvious thing' to be negated in the future. Is such a step possible?

4. Critique of Axiomatic Approach

It is a specific status of quantum axiomatics that it should reflect phenomenology. In order to verify the phenomenological background of quantum logical axioms a careful identification must be made in order to specify the elements of physical reality which correspond to the abstract 'questions'. At this point axiomatic theory is elusively elegant. A 'question' ('proposition'), we say, is an arbitrary macroscopic arrangement which, when interacting with a micro-object, may or may not produce a certain definite macroscopic effect: the presence of the effect is conventionally taken as the answer 'yes' whereas its absence is 'no' (or vice versa). Now, it is argued, the validity of the basic axioms of quantum logic (apart from weak modularity) is almost a matter of tautology. For instance, two 'yes-no measurements' with the identical 'certainly yes' domains are obviously testing for the same feature, and so the difference between them is not essential; this motivates the identity law (2.1). Similarly, the existence of a unique orthocomplement a' for an arbitrary 'yes-no' arrangement a is beyond discussion: a' is simply that some measuring arrangement with the roles of 'yes' and 'no' interchanged. An apparently more involved problem concerns the existence of the union $a \vee b$ and intersection $a \wedge b$ for any $a, b \in Q$. Here, some plausible existence arguments can also be given, through the constructive prescription is not clear. The above arguments would be indeed difficult to reject if not for the circumstance that the underlying definition is oversimplified. In spite of its elegant generality, the idea of a 'question' as a quite arbitrary macroscopic arrangement which produces a certain macroscopic alternative effect is wrong. To illustrate this, consider a statistical ensemble of any objects and a macroscopic device which yields the answer 'yes' for an average of $\frac{1}{2}$ of them in a completely random way. A good approximation is a semitransparent mirror in the path of a photon beam (Figure 1).

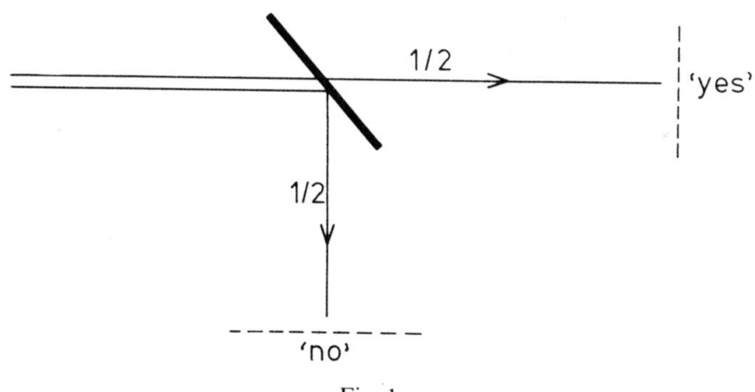

Fig. 1.

No doubt, this is a certain macroscopic arrangement producing a macroscopic alternative effect: either the photon reaches the screen 'yes' or it does not. However, the arrangement on Figure 1 cannot be considered one of 'questions'. If it were, it would produce a sequence of catastrophes in the structure of 'quantum logic'. First of all, it would not be clear which device is the 'negative' of the semitransparent mirror a. By insisting on the purely verbal solution (just the interchange of 'yes' and 'no') one would conclude that a' is acting, in fact, identically as a: for it too gives the answer 'yes' in a completely random way for an average of $\frac{1}{2}$ of the beam photons. Thus, $a' = a$. This would further imply: $0 = a \wedge a' =$

$= a \wedge a = a = a \vee a = a \vee a' = \mathrm{I}$ and so, the whole structure of Q would collapse.

One might reply, that the axioms of quantum logic are exact, but they must be properly understood. The semi-transparent mirror in Figure 1 is not a good example of a 'question' since it is not at all a measuring device: it does not verify any physical property of the transmitted photons. This is a good answer, but it means that the whole approach of 'quantum logic' should start from an information which is inverse to the usually given. *Not every arrangement producing a macroscopic alternative effect is a question (yes-no measurement)* – the right information should read. Indeed, one feels, that in order to be a quantum mechanical measuring device, the macroscopic arrangement should do something more specific than merely produce the 'yes' and 'no' effects in an arbitrary way. In some axiomatic approaches this is assured by requiring that the 'yes-no measurement' should have the non-trivial certainty domains: there should be some microsystems for which the answer 'yes' is certain and some other for which the answer 'no' is certain. This requirement eliminates the semi-transparent window as an element of Q. However, it is still far from sufficient. To see that, it is enough to consider two hypothetical macroscopic devices A and B acting on mixtures of red, yellow and violet light. The device A transmits the red photons and absorbs the yellow and violet ones: however, it re-emits on average $\frac{1}{2}$ of the absorbed yellow photons in form of red photons. The device B is also transparent for the red photons and absorbs the yellow and violet ones: now, however, $\frac{1}{2}$ of the violet photons are re-emitted in form of red photons. Schematically:

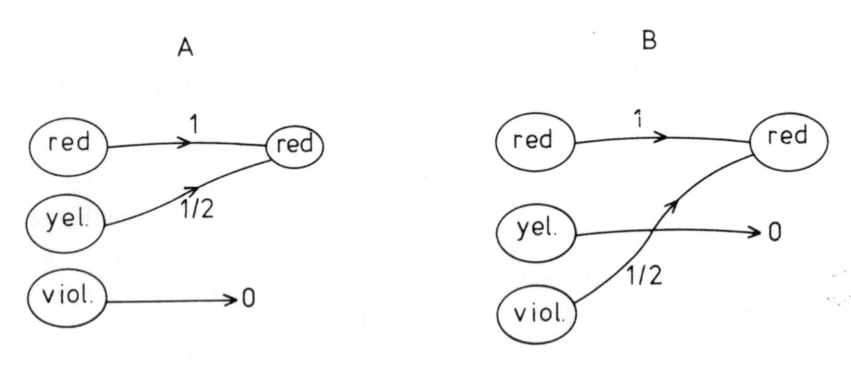

Fig. 2.

Both devices A and B have a common 'certainly yes' domain: they are completely transparent only to the red photons. They also have the property of performing idempotent operations on photon mixtures which in some treatments is considered a fundamental quality of the 'yes-no measurements'. However, A and B have different domains of 'certainly no' and so, they are not physically equivalent. This difference, in spite of Axiom I, cannot be considered non-essential and absorbed into the identity relation (2.1). Indeed, if we insisted that A and B are merely two different physical realizations of that same abstract question $a \in Q$, we would have two essentially different prescriptions for production of the negative a': once by taking A' ('certainly yes' for the violet) and once by taking B' ('certainly yes' for the yellow). Formally: $A = B$ but $A' \neq B'$. In consequence, something would be broken in the assumed structure of Q: either the identity axiom (2.1) or the uniqueness of the orthocomplement. We therefore reach the conclusion that the macroscopic devices A and B are still not 'good enough' to represent the abstract 'questions'. Indeed, the most essential condition is still missing. According to Ludwig's thermodynamical condition [5] the macroscopic 'yes-no measuring device' apart from possessing non-trivial certainty domains must also have the property of minimizing the randomness of the 'yes' and 'no' answers. A generalized version of this idea was employed in [6] by requiring that, for a given 'certainly yes' domain, the yes-no measurement should have a maximal possible 'certainly not' domain. This requirement is, finally, the sufficient condition which allows one to distinguish the subclass of those macroscopic devices which correspond to the abstract 'questions'. An essential problem now arises: is it necessarily so, that the counter-examples against the quantum logical axioms must automatically vanish when the class of the macroscopic 'yes-no' arrangements is restricted to the subclass of proper random-minimizing 'yes-no measurements'?

If the orthodox theory is not a priori assumed, the answer to this question must remain conditional. It depends essentially on the validity of a certain intuitive image which we usually associate with the phenomenology of physical objects and which, in general, may or may not be true. According to this image, each 'question' $a \in Q$ determines a certain specific property of micro-objects: the objects having that property are those for which the answer 'yes' is certain. Now, we intuitively assume that for each domain of micro-objects which possess a certain 'property' there is a unique complementing domain of micro-objects with an 'opposite property': so that, once it is known for which objects the answer of the 'yes-no measurement' is 'certainly yes' it is also uniquely determined for which ones it should be 'certainly no'. This image is true in orthodox quantum mechanics because of the orthogonal structure of the closed vector subspaces in a Hilbert space. However, it may be not of universal validity. In fact, it is not a logical impossibility to imagine a hypothetical physical world where to every domain of micro-objects with a certain special property there would be many possible 'complementing domains' corresponding to many possible ways of being 'opposite' to that property. If that were so, there could exist many random minimizing 'yes-no measurements' with a common domain of 'certainly yes' and different 'certainly no' domains. A hypothetical sequence of such devices is represented in Figure 3.

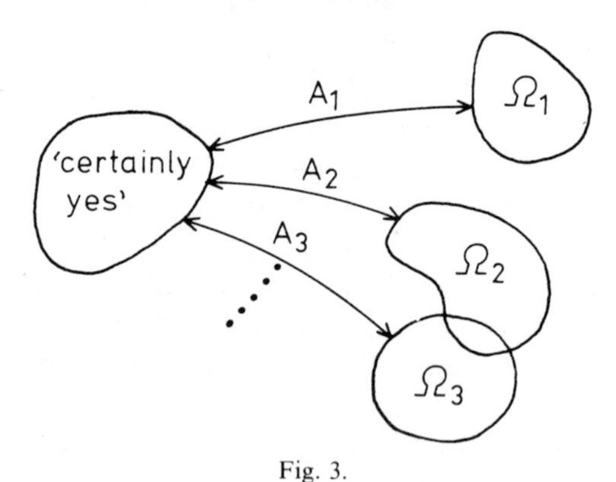

Fig. 3.

The devices $A_1, A_2, \ldots$ schematically represented in Figure 3 choose the same domain of micro-objects on which the answer should be 'certainly yes' but they minimize the randomness in favor of various 'certainly no' domains $\Omega_1, \Omega_2, \ldots$ For each of those devices the verbal negation operation (yes $\leftrightarrows$ no) could be easily performed leading to a sequence of devices $A'_1, A'_2, \ldots$ with different 'certainly yes' domains $\Omega_1, \Omega_2, \ldots$ Contrary to Axiom I, the devices $A_1, A_2, \ldots$ would be physically different, and even if we tried to neglect the difference by insisting that (2.1) defines the right physical equivalence, the negatives $A'_1, A'_2, \ldots$ could no longer be identified on that same principle. It thus becomes clear that the axioms of 'quantum logic' are not so absolute as they seem at the first sight. Even the apparently obvious laws of identification (2.1) and orthocomplementation (2.2–5) are not logically inevitable. Similarly, like the distributive law of classical logic, they are conditioned by the physical properties of the corresponding micro-objects. This suggests that before deciding what the 'quantum logic' is and which axioms it must fulfill, the theory should go deeper and look for the justification for the axiomatic structures in physics of the statistical ensembles themselves. The steps taken in this direction lead to the recently formulated convex scheme of quantum mechanics.

5. Convex scheme of q.m.

In the orthodox approach to quantum logic the statistical ensembles are an implicit counterpart. Their fundamental role has been rediscovered in the convex scheme of quantum mechanics [1, 5, 6]. The basic concept of this scheme is that of a *quantum state.* Given a statistical ensemble of certain micro-objects the *state* stands for an averaged quality of a random ensemble individual. Formally, states are equivalence classes of statistical ensembles. Given micro-objects of certain definite kind (e.g. electrons) the fundamental structure of the scheme is the set S of all states. If the micro-objects obey orthodox quantum mechanics, then S is the set of all positive operators with unit trace in a certain Hilbert space $\mathscr{H}$ ('density matrices'). If it is not a priori assumed that the orthodox theory holds, it may only be granted that S has a structure of a convex set: the convex combinations $p_1x_1+p_2x_2$ for $x_1, x_2 \in S$ and $p_1, p_2 \geqslant 0, p_1+p_2=1$ mean the state mixtures and the extremal points of S represent the pure states. In principle, S might be considered a convex set 'in itself' with the convex combination axiomatically introduced [3]. However, for the sake of illustrative qualities, one usually represents S as being embedded in a certain affine topological space E which can be constructed by a formal extension of S: the points in S then represent the pure and mixed states of the system, whereas the points of E out of S have no physical interpretation [6]. For the reason of physical completeness it is assumed that S is a closed convex subset of E.

Though the structure of S reflects a relatively simple phenomenology (it only shows which states are the mixtures of which other states) there is an extensive physical information contained in the geometry of S. In particular, shape of S determines the structure of the macroscopic alternative measurements which is so fundamental in other axiomatic approaches. This is due to the following concept of a normal functional.

DEFINITION. Given an affine space E with an affine linear combination $\lambda_1 x_1 + \lambda_2 x_2 (x_1, x_2 \in E, \lambda_1, \lambda_2 \in R, \lambda_1 + \lambda_2 = 1)$ a function $\phi: E \to R$ is called *linear* if $\phi(\lambda_1 x_1 + \lambda_2 x_2) = \lambda_1 \phi(x_1) + \lambda_2 \phi(x_2)$ for every $x_1, x_2 \in E, \lambda_1, \lambda_2 \in R$, $\lambda_1 + \lambda_2 = 1$. Given an affine topological space E and a closed convex subset $S \subset E$, a continuous linear functional $\phi: E \to R$ is called *normal* on S iff $0 \leqslant \phi(x) \leqslant 1$ for every $x \in S$.

The normal functional admit a simple geometric representation. Any non-trivial linear functional in E can be represented by a pair of parallel hyperplanes on which it takes the values 0 and 1. Now, the functional ϕ is normal on S if the subset S is contained in the closed region of E limited by the hyperplanes $\phi = 0$ and $\phi = 1$.

The normal functionals have a natural physical interpretation. Let $x \in S$ be a statistical ensemble and suppose, that there is a macroscopic device which produces a certain macroscopic alternative effect 'yes-no'. If one translates the 'yes' and 'no' into the numbers: 'yes' = 1 and 'no' = 0, the action of the device is completely characterized by a number $\phi(x)$ $(0 \leqslant \phi \times \times (x) \leqslant 1)$ which represents the statistically averaged answer to the individuals of the ensemble x. Since it is implicit in the definition of the statistical ensemble that it is composed of independent individuals, the process of testing any mixed ensemble is equivalent to testing independently each of the mixture components. Consistently, $\phi(p_1 x_1 + p_2 x_2) = = p_1 \phi(x_1) + p_2 \phi(x_2)$ and so, every 'yes-no' arrangement defines a certain normal functional on S. Here, no limitations are present which are essential for the 'quantum logic'. Every macroscopic alternative arrangement is included in the scheme and is mathematically represented by a normal functional, no matter whether or not it minimalizes the random element in the 'yes' and 'no' answers. We thus reach a generalized scheme of quantum theory based on the theory of convex sets ('convex scheme' [6]). In that scheme the collection of all states of a physical system is represented by a closed convex set S in an affine topological space. The set of all macroscopic 'yes-no' devices corresponds to the collection of all normal functionals on S mathematically represented by all possible ordered pairs of closed hyperplanes enclosing the set S and labelled by numbers 0 and 1 (see Figure 4).

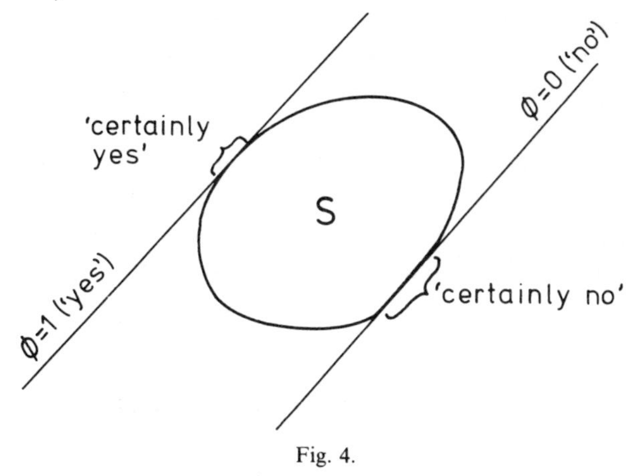

Fig. 4.

Since the set of the normal functionals is determined by the shape of the convex set S, so is also the collection of macroscopic 'yes-no devices'. Consequently it is a feature of the convex scheme that in it the structure of the 'yes-no measurements' is not decreed a priori but is determined by the more fundamental structure of the statistical ensembles. By analyzing the precise mechanism of this determination we reach a certain new structure which is a natural candidate for a replacement of traditional 'quantum logic'.

6. Logic of properties

It is a controversial problem, whether the formalism of quantum theory can be used to describe the properties of the single micro-object 'as it is, in all its complexity' (Piron [7]). The single act of measurement in quantum mechanics is not conclusive, and therefore, the direct interpretation of quantum mechanical formalism is that of a statistical scheme. The notion of property of a single system can however be introduced as a next abstraction stage of the theory. In the axiomatic approach of Jauch and Piron [4, 6] this is done by analyzing the structure of Q. Below, another method will be employed which departs directly from the properties of statistical ensembles.

Statistical ensembles are, in a way, macroscopic entities: though it might be impossible to predict the behaviour of a single micro-individual in a given physical situation, one can predict the behaviour of the ensemble as a whole. Therefore, there is no difficulty in defining the physical properties of the ensembles. By saying that a certain ensemble has a certain property we simply have in mind that the ensemble behaves in a specified way in some definite physical circumstances. If now the ensembles are represented by points of the convex set S, the properties are just the subsets of S. It is still an open question, whether a subset of S should fulfill some regularity requirements (such as the measurability) in order to represent a physically verifiable property. As pointed out by Giles [9] the answer must depend upon the degree of idealization which is permitted by the theory.

The main difficulty with the single individuals in a statistical theory lies in the fact that there is no immediate correspondence between the properties of the ensembles and the properties of the individuals. In fact, not every property of the ensemble is of such a nature that it may be attributed to each single ensemble individual. A strictly macroscopic example is obtained by considering a human ensemble composed half of men and half of women: the fifty-fifty composition then is a property of the ensemble which, however, cannot be attributed to each single ensemble individual. Quite similarly, one can have a beam of photons of which the average fraction $\frac{1}{2}$ penetrates through a certain Nicol prism. However, it may be that the ability of penetrating through the prism with the probability $\frac{1}{2}$ cannot be attributed to each single beam photon, for the beam is just a mixture of two types of photons one of which is certainly transmitted and the other certainly absorbed by the prism. In general, a property P of statistical ensemble is a proper starting point for a definition of a certain property of the single micro-objects if two conditions hold: (1) whenever two ensembles have the property P their mixtures must also have it, and (2) whenever a mixture has the property P, each of the mixture components must also have it. These requirements mean that the properties of the single microsystems are represented only by special subsets $P \subset S$ which fulfill the following definition [6].

DEFINITION. Given a convex S, a *wall* (also: *face*) of S is any subset $P \subset S$ such that: (1) $x_1, x_2 \in P$, $p_1, p_2 \geqslant 0$, $p_1 + p_2 = 1 \Rightarrow p_1 x_1 + p_2 x_2 \in P$, and

(2) $p_1x_1 + p_2x_2 \in P$ with $x_1, x_2 \in S$, $p_1, p_2 > 0$, $p_1 + p_2 = 1 \Rightarrow x_1, x_2 \in P$. Geometrically, a *wall* is any convex subset of S which possesses the property of 'absorbing intervals': whenever P contains any internal point of a certain straight line interval $I \subset S$ it must also contain the whole interval I.

The concept of a wall generalizes that of an extreme point: the extreme points are just one-point walls of S. Any non-empty convex set S has two improper walls: the whole of S and the empty set $\emptyset$. For any convex set the walls form a partially ordered set with the ordering relation $\leqslant$ being the set theoretical inclusion $\subset$. As seen from the definition, the common part of any family of walls is also a wall. This implies that the walls form a lattice: for any two walls $P, R \subset S$ the greatest lower bound $P \wedge R$ is just the common part $P \cap R$ whereas the lowest upper bound $P \vee R$ is obtained by taking the common part of all walls containing both P and R. If the points of S represent the pure and mixed states of a certain hypothetical system, the walls of S represent the possible properties of the system ordered according to their generality. In particular, the whole of S represents the most general property possible (no property) whereas the empty wall $\emptyset$ stands for the impossible property (no system with that property). The extreme points of S (if they exist) are atoms in the lattice of walls: they correspond to maximally specified properties, in agreement with the Dirac idea of pure states as being the maximum sets of non-contradictory information which one can have about the microsystem. It still remains an open question what sort of regularity requirements a wall should fulfill in order to be an operationally verifiable property. The standard quantum mechanical convention is to consider the lattice of closed walls of S as representing the physically essential properties of the system; this lattice will in future be denoted by $\mathbb{P}$.

The existence of normal functions on S allows one to define a natural notion of orthogonality in $\mathbb{P}$.

DEFINITION. Two properties P, R are called *excluding* or *orthogonal* ($P \perp R$) if there is at least one macroscopic 'yes-no' arrangement which answers certainly 'yes' for systems with the property P and certainly 'no' for the systems with property R (or vice versa). Thus, $P \perp R$ if there is a macroscopic device able to distinguish the property P from the property R without an element of probabilistic uncertainty.

The set of properties $\mathbb{P}$ with the relations of inclusion ($\leqslant$) and exclusion

($\perp$) is that structure of quantum theory which most directly reflects the nature of micro-objects. It has been thus proposed that the lattice $\mathbb{P}$ should be considered the 'logic' of a quantum system instead of the lattice of macroscopic measurements Q [6]. The above idea of quantum logic is wider than the orthodox one. The 'propositions' (properties) here are not necessarily in one-to-one correspondence with some 'yes-no measurements'. The 'property' is an abstracted quality of statistical ensembles and therefore, it should be verifiable: however, it is not a priori supposed that the verification might be always reduced to a single act of measurement. In spite of a more abstract sense of $\mathbb{P}$, the problem of the validity of the standard lattice theoretical axioms becomes much simpler for the 'logic of properties'. In fact, the identity axiom (2.1) is automatically fulfilled, since the partial ordering $\leqslant$ is now the set 'theoretical inclusion'. The lattice axiom, too, automatically holds because the closed walls of S must form a lattice. It is not so with the orthocomplementation law which has now a quite different status. The notion of 'negation' is not immanent for the properties. What becomes natural here is the more primitive relation of the exclusion $\perp$. Depending on the structure of that relation the operation of negation can or cannot be constructed on $\mathbb{P}$. The following definition seems to express the physical idea of what the negation is.

DEFINITION. Let P be a property and let $P^\perp$ denote the subset of all properties which are orthogonal to P. If in $P^\perp$ a greatest element exists,

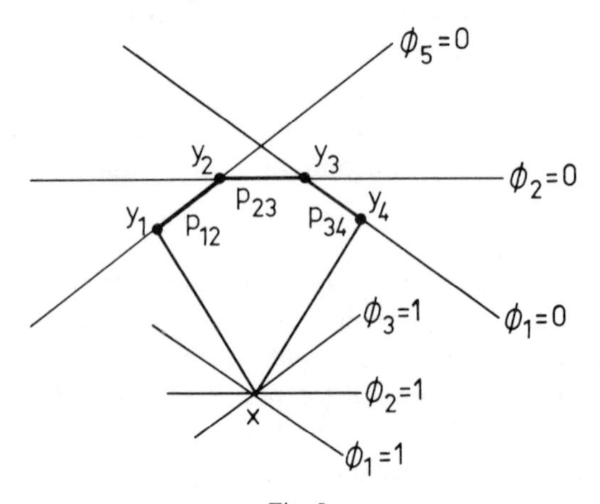

Fig. 5.

this element is called the *negative* of P and denoted P'. If for every $P \in \mathbb{P}$ the negative P' exists, we say that the logic $\mathbb{P}$ admits negation.

As is easily seen, the existence of negation, in general, is not ensured by the structure of the walls. A hypothetical case where negation could not be constructed because of the geometry of S is shown on Figure 5.

For the convex set represented here the subset composed of one point x is a wall and the family $\{x\}^{\perp}$ contains the seven non-trivial walls: the four one-point walls $\{y_1\}$, $\{y_2\}$, $\{y_3\}$, $\{y_4\}$ and the three straight line segments P_{12}, P_{23}, P_{34}. The family $\{x\}^{\perp}$ thus contains three maximal walls P_{12}, P_{23}, P_{34} but it does not contain the greatest one: the convex set S does not possess a wall which would contain the three segments P_{12}, P_{23}, P_{34} and be orthogonal to x. As a consequence, no unique orthogonal complement can be defined for the 'property' $P=\{x\}$. The above situation has not very much to do with the possibility of interchanging the 'yes' and 'no' answers in the 'yes-no measurements' and cannot be excluded by considering the nature of the macroscopic measuring devices. Inversely, this is the absence of the situations like that represented in Figure 5 which must be first granted to explain the origin of the usually assumed structure of Q. In fact, if the convex set in Figure 5 represented the collection of all pure and mixed states of a certain physical system, the structure of the 'properties' would make possible the existence of three different 'yes-no' measurements with the same domain of 'certainly yes' (the pure state x) which would, however, minimize the random element in different ways, by choosing three different 'certainly not' domains P_{12}, P_{23}, P_{34}. This would lead to a non-orthodox structure of Q with the identity law broken (see also [6]). This shows that the logic of properties $\mathbb{P}$, in a sense, lies one level deeper than the phenomenology of 'yes-no measurements'. The existence or non-existence of negation in $\mathbb{P}$ is one simple fact which justifies or disproves the whole system of axioms which are traditionally employed to describe the structure of 'questions'. An essential problem now arises: have we indeed some universal reasons to believe in the orthocomplemented structure of $\mathbb{P}$?

If we dismiss some verbal arguments, we are really left with one intuitive picture which can be of importance. This is the picture of matter and of a certain selection process which subtracts a component of matter with a certain definite property. Now, if the subtraction is done with enough care, the component which remains ('the rest') depends only on *what* was subtracted but it does not depend on *how* it was subtracted. This simple picture is, in fact, one of the deepest constructional principles of present day theory and is the true origin of the subsequent image of ℙ as being an orthocomplemented lattice. Indeed, one usually takes for granted that each 'property' distinguishes a certain component of matter: 'the rest' then uniquely defines the 'complementing property'. This idea finds a particular realization in classical theory, where the properties correspond to subsets in a classical phase space and the 'subtraction' is the operation of taking the set theoretical complement. A different mechanism stands for the same in orthodox quantum theory. Here the states of matter are described by vectors in a linear space (wave vectors) which obey a linear evolution equation. Now, if a wave vector is selected by subtracting a certain component, the uniqueness of 'the rest' is due to the existence of the linear operations. This explains the strong position of the orthocomplementation axiom in the present day theory: whenever one deals with some quanta which are well described by a linear wave equation, the orthocomplemented structure of 'properties' will naturally appear. On the other hand, this also indicates that the orthocomplemented structure of ℙ might not be universal: it does not express the nature of *any* theory, it just expresses the essence of linearity. Is linearity a necessary attribute of quantum mechanics? In spite of the traditional philosophy of the superposition principle, schemes based on non-linear wave mechanics have always been a tempting alternative for quantum theory. One might expect them to contribute something to the understanding of the measurement problem: for a possibility is open that the Schrödinger evolution equation and the measurement axioms are just two opposite approximations to a still undiscovered theory. Thus, the quantum axiomatics should not be too quickly closed. This may become of special importance in problems which involve the quantization of gravity.

In fact, if the gravitational field has a quantum character, an intriguing problem concerns the behaviour of a hypothetical single graviton. Is this behaviour similar to the dynamics of the macroscopic gravitational universe governed by Einstein's equations? In principle, it must not be so. It is possible that the single graviton in vacuum (if such an entity exists) is well described by a certain linear law, in agreement with the spirit of orthodox quantum mechanics, and that the non-linear behaviour of the macroscopic gravitational field is a secondary phenomenon due to interactions in a cloud of many gravitons. In this case the formalism of operator fields in Hilbert spaces would be sufficient to describe the quantum gravitation. However, this hypothesis has some disadvantages. In fact, if the graviton were described by a linear wave, a question would arise, as to in what space-time this wave propagates? Is it just the flat Minkowski space-time? If so, general relativity would be a theory with a background of Minkowski metric masked by clouds of gravitons. This is, however, against an innate aesthetics of general relativity, where the background metric which is not seen under the cover of the macroscopic field is unphysical and should not enter into the formulation of the theory. Thus, it may be that the situation is different. Perhaps, even a single graviton in vacuum modifies the space time in which it exists and so, creates a nonlinearity in its own propagation law. This would lead to a new picture of quantum theory where the selection processes could no longer be associated with linear decomposition operations of certain 'state vectors' and the mechanism which had accounted for the uniqueness of the orthocomplement in the orthodox theory would no longer be valid. The resulting properties of 'non-linear quanta' would not have to imitate the orthogonality structure of the closed vector subspaces in $\mathscr{H}$, but they could form a generalized type of logic where no unique negation operation can be constructed.

Warsaw University

BIBLIOGRAPHY

[1] Davies, E. B. and Lewis, J. T., 'An Axiomatic Approach to Quantum Probability', *Commun. Math. Phys.* **17** (1970), 239.
[2] Finkelstein, D., *The Logic of Quantum Physics*, N.Y. Acad. Sci., 1963, p. 621.
[3] Gudder, S., 'Convex Structures and Operational Quantum Mechanics', *Commun. Math. Phys.* **29** (1973), 249.
[4] Jauch, J. M. and Piron, C., 'On the Structure of Quantal Proposition Systems', *Helv. Phys. Acta* **42** (1969), 842.
[5] Ludwig, G., 'Versuch einer axiomatischen Grundlegung der Quantenmechanik und allgemeinerer physikalischer Theorien', *Z. Naturforsch.* **22a** (1967), 1303; **22a** (1967), 1324.
[6] Mielnik, B., 'Theory of Filters', *Commun. Math. Phys.* **15** (1969), 1; 'Generalized Quantum Mechanics' (to appear in *Commun. Math. Phys.*).
[7] Piron, C., 'Axiomatique quantique', *Helv. Phys. Acta* **37** (1964), 439.
[8] Pool, J. C. T., 'Semi-Modularity and the Logic of Quantum Mechanics', *Commun. Math. Phys.* **9** (1968), 212.
[9] Giles, R., 'Axiomatics of Quantum Mechanics', *J. Math. Phys.* **11** (1970), 2139.

Z. MARIĆ AND DJ. ŽIVANOVIĆ

THE STOCHASTIC INTERPRETATION OF QUANTUM MECHANICS AND THE THEORY OF MEASUREMENT

There is by now a variety of views concerning the general position of quantum physics in natural philosophy. This variety appears clearly in measurement theory, which is a necessary part of any consistent description of physical phenomena on the quantum level. The orthodox view reflects a radical departure from the traditional attitude in the analysis of the external world, an attitude of mind inherited from the rationalistic tradition. Subjective elements have been introduced into physics with the von Neumann paradigm, and this seems to be unavoidable within the orthodox scheme. Another variant of this scheme, i.e. Bohr's analysis of the measurement process, requires the abandonment of any description of nature that would be unique and independent of different 'levels' of phenomena. It is no wonder that alternative interpretations have been sought, since the orthodox view implies philosophical consequences which are not easy to accept.

In measurement theory the central dilemma opens with Bohr's insistence on the 'wholeness' of the system and the apparatus during the measurement. But the term 'wholeness' does not unambiguously determine by itself either the ontological attitude one can take, or the manner in which it should be incorporated into the physical theory. In both these aspects we are left with at least two possibilities.

Philosophically, one may consider measurement as something fundamental, as an ultimate source of information. The systematic knowledge of the phenomena, i.e. the physical theory, is then nothing else but another expression for the correlated data. Or, on the contrary, and in concordance with the secular scientific tradition, we may consider measurement as a physical process and analyze this process by specifying its conditions in any concrete physical situation.

Physically, the 'wholeness' may be conceived as something nonanalyzable, something which should be present only through its manifestations (the complementarity, the impossibility of a unique reduction of a quantum ensemble into pure state sub-ensembles, etc.). Or, on the con-

M. Flato et al. (eds.), Quantum Mechanics, Determinism, Causality, and Particles, 137–146. *All Rights Reserved.*
Copyright © 1976 by D. Reidel Publiishing Company, Dordrecht-Holland

trary, we may insist that it should be analyzable, even, if necessary, by using the entities which are abstract in the sense of their being unobservable. This alternative is known as the search for a hidden variable theory, the term referring to the aim of understanding the quantum probability features in analogy with the probability theory as used in the classical mechanics. Another way of stating the above-mentioned alternatives is to speak about the complete description of a physical system by the wave function ψ, or to deny the completeness of this description, as was suggested by de Broglie, Einstein and Schrödinger.

In their 'primitive' form, hidden variable theories were excluded long ago by the von Neumann theorem, but in a more sophisticated version, which explicitly takes into account the measurement process, hidden variable theories are possible. One such theory was proposed by de Broglie as early as 1927 [1] and further developed by Bohm much later (1952) [2]: the pilot-wave theory. The von Neumann theorem is not relevant here because two incompatible measurements produce two different ensembles; this reveals the fact that the hidden variables of the measuring apparatus are as important as those of the measured system. The invention of such a model is of great importance. It opens a way towards a more profound theory (e.g. the theory of the double solution) and at the same time, it demonstrates that the prevailing theory of quantum phenomena with the corresponding interpretation of its mathematical formalism is not the only possible one. The result was achieved by giving the ψ-function a double role: it reflects the probabilistic features and at the same time it is considered as a physically real field. The payoff comes in Bohm's measurement theory: the reality ascribed to the ψ-field mysteriously disappears once the reading on the scale of the apparatus is performed. The ψ-field is not subjective in the von Neumann sense, but nevertheless it seems to incorporate elements of the previous knowledge into the physical reality (we note that the physical reality we are speaking about is the physical reality *strictu senso* and not the image of it which we can form). However, detailed analysis of the measurement process has revealed a feature which might be the cornerstone of the future theory. In order to describe mathematically the interaction process between two physical systems, one of which plays the role of the apparatus, something which could express mathematically the nonlocality of the interaction and the impossibility of analyzing the system in its complete autonomy is needed, rationalizing in such a way the 'wholeness' of physical processes at the quantum level in agreement with the Bell theorem.

With this in mind, one is in front of a no-man's-land, where many possibilities are open. One way might be to take seriously the connection between the Schrödinger equation and the equations for the Brownian motion of a particle [3], and to start directly with the dynamic equations relating the stochastic quantities which correspond to the position and the velocity in the classical limit. By the use of stochastic variables one avoids the necessity of explaining, at this early stage of the theory, the probabilistic character of quantum laws. The underlying physical picture is that of a perpetual motion of the ultimate constituents of matter, a kind of 'Zitterbewegung' which, although undetectable, might be taken as the starting point in the elaboration of the description of physical phenomena. Some people might prefer to see in it just a methodological element, or perhaps a manifestation of intercorrelations in the underlying physical substratum.

The stochastic theory of this type already exists in a preliminary form. Weizel [4], Nelson [3, 5] and La Peña-Auerbach [6] have been able to show the equivalence between the non-relativistic Schrödinger equation and theory of the frictionless Brownian motion with the diffusion coefficient proportional to Planck's constant. The problem of the spin and that of the identical particles has been also successfully treated in the non-relativistic approximation [7, 8].

This particular approach has some features which make its further investigation worthwhile. Instead of starting from the Schrödinger equation as a fundamental postulate and then simply reinterpreting it, this equation is derived from postulates which are inherited from the logical structure of classical physics. But in order to be complete, in the sense of being completely equivalent to the standard quantum mechanics, the theory should have its own theory of measurement. This problem has not attracted a great deal of attention (with the exception of Ref. [3] and [5]) and we wish to make a comment in this respect.

In this theory the dynamics of a system consisting of N particles is described by a stochastic process $X(t)$ on the configuration space R^{3N}; $X(t)=\{X_{i\alpha}(t)\}$; $i=1,\dots, N$; $\alpha=1, 2, 3$. The displacement during any finite time interval is given by the integral over infinitesimal Gaussian increments $\mathrm{d}X(t)$ with constant variance. The mean is a function of $X(t)$ [9].

One has:

$$dX(t) = C[t, X(t)]\,dt + dW(t). \tag{1}$$

The Wiener process $dW(t)$ is characterized by:

$$\begin{aligned} &E\,dW(t) = 0 \\ &E\,dW^T(t)\,dW(t) = \sigma^2\,dt, \end{aligned} \tag{2}$$

where E denotes the expectation and the superscript T refers to the matrix transposition.

The mean forward velocity C and the diffusion matrix σ^2 are defined in terms of conditional expectations with respect to the past state:

$$C[t, X(t)] = \lim_{\Delta t \to 0+} E\left\{\frac{X(t+\Delta t) - X(t)}{\Delta t}\,\middle|\, X(t)\right\} \equiv DX(t) \tag{3}$$

$$\sigma^2[t, X(t)] = \lim_{\Delta t \to 0+} E\left\{\frac{[X(t+\Delta t) - X(t)]^T[X(t+\Delta t) - X(t)]}{\Delta t}\,\middle|\, X(t)\right\} =$$

$$= \left\| \frac{\hbar}{m_i}\,\delta_{ij}\delta_{\alpha\beta} \right\|; \quad i, j = 1, 2, \ldots, N; \quad \alpha, \beta = 1, 2, 3. \tag{4}$$

Here, m_i denotes the mass of the ith particle and $\hbar$ is Planck's constant divided by 2π. As it stands, it is connected with the definition of the diffusion parameter.

Almost all trajectories $X(t)$ are continuous, but they are not differentiable. Therefore, the velocity does not exist. The forward derivative D, defined in (3) can be extended to any function of $X(t)$ by means of a Taylor series expansion.

The image of the physical process which follows consists of an independent diffusion process for each particle but the mean (forward) velocity of any particle depends on the position of all others. The theory is strongly non-local and the configuration-space description becomes necessary.

The transition probability $P(Xt \mid X't')$, characterizing the Markovian process, can be found from the forward Fokker-Planck equation. All what is needed is to know C and σ^2 in R^{3N+1} space (coordinates plus time). Then, the probability density $\rho(X, t)$ is the solution of the equation:

$$\frac{\partial \rho}{\partial t}+\nabla(\rho C)-\frac{\sigma^2}{2}\nabla^2\rho=0. \tag{5}$$

In the same manner the transition probability $P(X, t \mid X', t')$ can be found from the backward Fokker-Planck equation in terms of the backward mean velocity C_* and the corresponding diffusion matrix σ_*^2. These two quantities are defined by using the conditional expectation with respect to the further state. From the general theory of Markov processes [3] it follows:

$$\sigma^2=\sigma_*^2 \tag{6}$$

$$C-C_*=\sigma^2\frac{\nabla\rho}{\rho}. \tag{7}$$

All such relations are essentially kinematic. In order to introduce dynamics, Nelson [3, 5] has defined the mean acceleration:

$$a[t, X(t)]=\tfrac{1}{2}(DD_*+D_*D)\,X(t). \tag{8}$$

This is a time-reversal invariant generalization of the second derivative.

The dynamic law is obtained by postulating

$$a[t, X(t)]=F[t, X(t)] \tag{9}$$

where $F\equiv\{f_{i\alpha}\}$; $i=1,\ldots, N$; $\alpha=1, 2, 3$) being the $3N$-dimensional vector built from the classical force (per unit mass) which acts on particles.

Using (7) we can eliminate C_* from (9) and obtain an equation which connects C and ρ. This equation together with (5) determines ρ and C for all times *if* the initial values $\rho(X, t_0)$ and $C(X, t_0)$ are known.

It was demonstrated that this system of equations is equivalent to Schrödinger's equation [5, 6]. Therefore, it follows, that there is one-to-one correspondence between quantum processes described by Schrödinger's equation and a certain class of Markovian processes. The probability density ρ of the stochastic mechanics is related to the Schrödinger ψ-function with the known relation: $\rho=|\psi|^2$.

It should be stressed, nevertheless, that the equivalence obtained is valid only for $\psi\neq0$ (except possibly at infinity). Therefore, all the stationary states but the ground state are excluded, since they all have the nodal surfaces. The problem appears also in other alternative theories, and the solution lies either in denying the physical relevance of these states [10], or in defining them as limits of the quasi-stationary states for which this difficulty does not arise [5]. (In measurement theory, where the quantum coupling between two systems determines the physical characteristic of the system considered, the problem might be circumvented, since the action of the apparatus necessarily destroys the nodal surfaces, as we shall see from further discussion.)

The stochastic mechanics, i.e. the system of equations for ρ and C ought to be completed by the measurement theory. This is also seen from the fact that the theory is incomplete as yet, since no rule is given for determination of the initial values $\rho(X, t_0)$ and $C(X, t_0)$. Unfortunately, at present we do not have a measurement theory expressed in terms of its own basic elements: the stochastic trajectory and the fundamental Newtonian-type dynamic scheme. Nevertheless, the established correspondence between stochastic mechanics and standard non-relativistic quantum mechanics enable us to explore tentatively different possibilities in order to see whether is it possible to think consistently of a measurement process in terms of the stochastic mechanics. In this respect, we might take one of the following attitudes.

(1) One might ignore the problem created by the von Neumann paradigm and speak about the disconnected series of preparations and measurements made on quantum ensembles [11]. This is not in contradiction with the minimal axiomatic scheme of quantum mechanics, although in some cases the delicate frontier between physics and semantics seems to play an exceptional role. This attitude is easily translated into the language of stochastic theory. Here, we speak of the action of a measuring apparatus that gives rise to a change of the Markovian process, i.e. during the measurement process the system passes from one Markovian process to another.

(2) It is possible to consider the 'wholeness' of the measurement process by including the position variables of the apparatus in the scheme. This line of thinking is two-sided: it might contain a bias towards an Everett-like interpretation; or, the apparatus coordinates may be considered as determining an external field of forces, different for different measurements. Then it follows immediately that two incompatible measurements are represented by two different Markovian processes. Contrary to the opinion that in this case the Brownian character of the apparatus sets a

limitation to the measurement precision [5], we now think that the apparatus is, from the very beginning, described by its macroscopic properties. In such case the difficulties do not lie in the overall Brownian motion of the apparatus, but in the impossibility of analyzing in detail the extremely complicated Brownian motion of the system during its interaction with the apparatus. The situation is even more complicated for the so-called 'measurement of the second kind'. As de Broglie has pointed out [12], the measurement can not always be effected by the separation of the wave function of the system into spatially separated wave packets and the subsequent detection of the particle by a mere establishment of its existence in one of the wave packets (this would be the so-called 'measurement of the first kind', which could be treated as the setting up of a new Markovian process in accordance with the new knowledge acquired). In a 'measurement of the second kind' the one-to-one correspondence is established between spatially separated wave packets of the 'indicator' particle and the states (not necessarily spatially separated) of another particle on which the measurement is being performed. The detection of the 'indicator' particle in one definite region of space results sometimes in its disappearance, and all the information one obtains refers to its coupling with the surviving particle, the subsequent behaviour of which has to be predicted. This could be a motivation for an analysis of the measurement process analogous to that of von Neumann in the orthodox theory, as well as the point at which the advocacy of measurement theory as a necessary part of the description of quantum phenomena finds its *raison d'être*.

(3) Therefore, let us consider the stochastic theory applied to the N-particle problem. For brevity's sake we shall speak only about two particles: $X(t)=\{X_i(t)\}$, $i=1, 2$. In this analysis we are guided by von Neumann measurement theory, which treats both the system *and* the apparatus as quantum objects. The stochastic variable $X_2(t)$ should be treated as the position of the 'indicator' particle, solving in such a way the above-mentioned difficulties concerning 'measurements of the second kind'. Nevertheless, we shall consider formally $X_2(t)$ as the 'apparatus coordinate', in order to be able to compare the measurement theory in the stochastic interpretation both with the orthodox measurement theory theory and with the Bohm's theory of measurement as given in his early papers [2].

By choosing a convenient interaction with an appropriate apparatus we can measure different 'observables' associated with the particle. The apparatus position $X_2(t)$ is a macroscopic variable (i.e. it is considered in the $m_2 \to \infty$ limit) and so the different values of an observable correspond to the *macroscopically distinct* 'readings'. We see that, just as in the case of the pilot-wave theory, these 'observables' do not belong to the particle alone, but are, in a sense, 'produced' through the interaction with an appropriate apparatus. But, with the one-to-one correspondence between possible (and directly observable) positions of the apparatus and the different states of the particle, it is, in principle, possible to determine the corresponding initial values of ρ_1 and C_1 after the 'reading' of the apparatus, defining in such a way the Markovian process which describes the further stochastic motion of the particle. The information obtained by the measurement therefore enables us to construct the new ensemble to which the observed particles belong *after* the measurement.

This analysis shows that the equivalence between two schemes holds only when the system is not subjected to the measurement. In order to make a stochastic process correspond to the act of measurement itself, we must consider the larger configuration space which includes the given apparatus too. This point is similar to the von Neumann measurement theory. But now the fact that the apparatus shows a determinate 'reading' in any single experiment is not ascribed to an 'action of the *abstract ego*' on the otherwise complete wave function, but it translates the circumstance that a stochastic theory can not describe an individual event and it is, therefore, by definition incomplete. The necessity of using a different configuration space for each possible apparatus seem to vindicate in some measure the concept of an 'unanalyzable wholeness'. In fact, it is a consequence of the non-local character of the theory. The non-locality makes the *undisturbed* observation of a system *impossible*.

One notes that the position $X_1(t)$ of the particle plays a special role, being the fundamental variable of the stochastic theory, but the stochastic trajectory cannot be experimentally detected, since it is also disturbed by the measurement. The stochastic theory should be considered as a 'hidden motion' theory rather than a 'hidden variable' one, since it does not demand (but also it does not exclude) an underlying determinism. It is a pure particle theory whose mechanistic character may not be acceptable to everyone, while it may be rather a point in its favor for others.

In conclusion we would like to point out that the stochastic interpretation is a logically consistent alternative theory of microscopic phenomena, in the nonrelativistic approximation. Nevertheless, it is not altogether free from difficulties similar to those which beset the orthodox interpretation.

First of all, in this case no clear criterion exists by which one can see that the physical system is a macroscopical one (except the $m \to \infty$ limit and that is not sufficient). So we cannot properly characterize the apparatus starting from the fundamental dynamical scheme. This circumstance might be due to the fact that we lack at present the proper language to describe the measurement process in terms of the stochastic theory. Lacking that, we can only use the equivalence between the stochastic interpretation and the orthodox one to construct the theory of measurement which is essentially a reinterpretation of the standard measurement theory.

Another difficulty which remains in the stochastic theory is that the non-locality (manifest in the observed existence of distant correlations) is postulated through the fundamental system of equations and not explained in terms of some more fundamental physical interaction between the particles. This non-locality plays an important role in the theory of measurement, being responsible for the 'wholeness' of the process. The difficulty could be resolved only if there were an underlying and more fundamental physical theory, the present stochastic theory being only a first step towards its formulation.

Nevertheless, in its preliminary form the theory may serve as a bridge between classical and quantum description of physical processes. It offers a better understanding of the relationship between methodological and objective elements of the physical theory by introducing the probability in the classical way. It does not pretend to give a complete description of an individual physical system and so it disposes of paradoxes such as the paradox of the Schrödinger's cat [3]. Further, the relation $\rho = |\psi|^2$ follows from the fundamental postulates of the stochastic theory and need not be separately demonstrated as, e.g., on the causal interpretation [10].

Finally, it might be supported by the possible existence of the fluctuating electromagnetic field [13], which might give rise to the stochastic motion of particles and at the same time be generated by this motion in some new kind of dynamical equilibrium.

However, it should be said that a fully relativistic stochastic theory does not exist at present, although some initial success in this direction has been achieved [14, 15].

Institute of Physics, Belgrade, Yugoslavia

BIBLIOGRAPHY

[1] de Broglie, L., *Compt. rend.* **183** (1926), 447; **184** (1927), 237; **185** (1927), 380.
[2] Bohm, D., *Phys. Rev.* **85** (1952), 166.
[3] Nelson, E., *Dynamical Theories of Brownian Motion*, Princeton University Press, Princeton, N.J., 1967.
[4] Weizel, W., *Z. Physik* **134** (1953), 264; **135** (1953), 270; **136** (1954), 582.
[5] Nelson, E., *Phys. Rev.* **150** (1966), 1079.
[6] de la Peña-Auerbach, L., *J. Math. Phys.* **10** (1969), 1620.
[7] de la Peña-Auerbach, L., *J. Math. Phys.* **12** (1971), 453.
[8] de la Peña-Auerbach, L. and Cetto, A. M., *Rev. Mex. Fis.* **18** (1969), 323.—
[9] Doob, J. L., *Stochastic Processes*, J. Wiley, London, 1953.
[10] Bohm, D. and Vigier, J. P., *Phys. Rev.* **96** (1954), 208.
[11] Margenau, H., *Ann. of Phys.* **23** (1963), 469.
[12] de Broglie, L., *La théorie de la mesure en mécanique ondulatoire*, Gauthier-Villars, Paris, 1957.
[13] Boyer, T. H., *Phys. Rev.* **182** (1969), 1374; **186** (1969), 1304; **D1** (1970), 1526; **D1** (1970), 2257.
[14] de la Peña-Auerbach, L., *Rev. Mex. Fis.* **19** (1970), 133.
[15] Rylov, Yu. A., *Ann. der Physik* **27** (1971), 1.

D. BLOKHINTSEV

STATISTICAL ENSEMBLES IN QUANTUM MECHANICS

1. Introduction

The interpretation of quantum mechanics presented in this paper is based on the concept of quantum ensembles. This concept differs essentially from the canonical one by that the interference of the observer into the state of a microscopic system is of no greater importance than in any other field of physics (see Refs. [1, 3]). Owing to this fact, the laws established by quantum mechanics are of not less objective character than the laws governing classical statistical mechanics.

The paradoxical nature of some statements of quantum mechanics which results from the interpretation of the wave function as the observer's notebook greatly stimulated the development of the idea presented below.

As early as in the initial stage of the development of quantum mechanics, many of these paradoxes were indicated by Professor de Broglie, to whom I am pleased to dedicate this paper on the occasion of his 80th anniversary.

2. Macroscopic setting

Quantum mechanics deals with the study of the behaviour of microscopic particles or their systems μ embedded in a definite macroscopic environment $\mathfrak{M}$. This environment (in what follows called macrosetting) is, in its essence, specified in a macroscopic manner, i.e. by indicating the location of macroscopic bodies, by specifying classcal fields, the temperature of the bodies etc.

It is obvious that measurements performed in this situation on a single specimen of the microscopic system yield no unambiguous information. Measurement repeated for the same macroscopic setting $\mathfrak{M}$ on a physically identical microsystem μ will, generally speaking, lead to another

M. Flato et al. (eds.), Quantum Mechanics, Determinism, Causality, and Particles, 147–158. *All Rights Reserved. Copyright © 1976 by D. Reidel Publishing Company, Dordrecht-Holland.*

random result. If such measurements are repeated an infinite number of times, and, as a result, there arises a definite reproducible distribution of the results, then we are concerned with a statistical ensemble. This ensemble consists of an infinite number $N(N\to\infty)$ of reproductions of the settings $\mathfrak{M}+\mu$.

An example of such an ensemble in classical physics is the Gibbs ensemble in which the macrosetting is given by a thermostat M_θ with definite temperature θ. The ensemble arises as a result of the reproduction of the settings $M_\theta+\mu$. The coupling between the molecular system μ and the thermostat is, in this case, assumed to be weak.

We note, that strictly speaking, the macrosetting $\mathfrak{M}$ comprises the whole set of macroscopic factors, including the measuring instrument Π and even the observer, which affect the microsystem.

When restricting ourselves to physical effects, we can completely delete the observer, and then the whole macrosetting will be kept within the framework of physics. Under certain conditions, we can divide this setting $\mathfrak{M}$ into two parts; the part M which determines the state of the microsystem irrespectively of the measuring instrument Π and the instrument itself:

$$\mathfrak{M}=M+\Pi \tag{1}$$

Within the framework of classical physics all the measuring instruments providing one with the full information on the configuration of a microsystem and its motion are, in principle, equivalent to one another, and thus form a unique class of instruments. This information reduces to indication of a point in the phase space $\mathscr{R}(p, q)$. Here by q we mean a complete set $q_1, q_2, \ldots, q_f$ of the generalized coordinates describing the space configuration of the system, and by p we mean a complete set $p_1, p_1, \ldots, p_f$ of the canonically conjugate momenta describing the motion of the system in question.

The distribution of the measurement results in a classical ensemble is given by the quantity:

$$\mathrm{d}W_\theta(p, q)=W_\theta(p, q)\,\mathrm{d}p\,\mathrm{d}q \tag{2}$$

It is assumed that in this case the macrosetting is defined by the temperature of the thermostat θ. The distribution (2) is complete, therefore all other distributions can be simply derived from it by making a transformation to other variables $(p'q')$.

It is worth noting that the distribution (2) contains both the characteristics of the macrosetting (temperature θ) and those of the microscopic system (dynamical variables p, q). This emphasizes that we speak of a microscopic particle embedded in a certain macrosetting rather than a microscopic particle in itself.

A classical measuring instrument may be arranged to be such that it will noticeably affect neither the system μ under measurement nor the macrosetting M. Therefore in classical ensembles, when describing the macrosetting we may exclude the measuring instrument so that $\mathfrak{M}=M$.

3. THE QUANTUM ENSEMBLE

The ensembles of quantum mechanics differ essentially from the classical ones. This difference can be formulated in its shortest form by means of the principle of complementarity. In the language of the theory of quantum ensembles, this principle reads: there exists no quantum ensembles in which both the mean square deviation $\overline{\Delta q_s^2}$ *of the coordinate* $q_s(s=1, 2, \ldots, f)$ and the one $\overline{\Delta p_s^2}$ of the momentum p_s conjugate to it are arbitrarily small. Mathematically this is expressed in the Heisenberg uncertainty relation

$$\overline{\Delta p_f^2}\,\overline{\Delta q_f^2} \geqslant \hbar^2/4 \tag{3}$$

According to this principle, all measuring instruments are divided into two sub-classes: space-time ones and momentumenergy ones. The instruments belonging to the first sub-class are capable of providing information on the complete set of the coordinate variables q and thereby indicating a point in the configuration space $\mathscr{R}(q)$. The instruments belonging to the second sub-class indicate a point in the momentum space $\mathscr{R}(p)$. It is impossible, in principle, to indicate a point in the phase space $\mathscr{R}(p, q)$ for the microsystem belonging to a quantum ensemble. By virtue of this division of instruments, we should in the case of a quantum ensemble indicate the measuring instrument Π, when formulating the macrosetting.

The macrosetting thus depends on the kind of the instrument, and this dependence can be represented symbolically as

$$\begin{aligned} \mathfrak{M}_p &= M + \Pi_p \\ \mathfrak{M}_q &= M + \Pi_q \\ \mathfrak{M}_a &= M + \Pi_a \\ &\text{-------} \end{aligned} \tag{4}$$

where by Π_p, Π_q, Π_a--- we mean the instruments measuring the dynamical variables p or q; or any others a---. It is seen from the scheme that we restrict ourselves to considering the cases when the part M of the macrosetting remains unaffected, in spite of the alteration of the instrument. This part forms the state of the microparticle in the ensemble. Following the Fock terminology, this is an instrument 'preparing' the initial state of a microparticle.

The suggested distinction between M and Π makes it possible to formulate the quantum ensemble in a more narrow sense, as an independent N times reproduction ($N \to \infty$) of the setting

$$M + \mu \tag{5}$$

Different measurements on the same particle μ are, generally speaking, incompatible. However, using an ensemble, it is possible to obtain any information about any simultaneously measurable variables if we divide N specimens of the ensemble into some parts N_p, N_q,...$N_a \to \infty$ in such a way that $N = N_p + N_q + \cdots + N_a$ and each N_p, N_q,...$N_a \to \infty$, and perform measurements with the aid of the instrument Π_p in the group N_p and with the aid of Π_q in the group N_q and so on.

In so doing, we find the distributions of the dynamical variables of the microsystem μ embedded in the macrosetting M. We denote these distributions as

$$\begin{aligned} \mathrm{d}W_M(p) &= w_M(p)\,\mathrm{d}p \\ \mathrm{d}W_M(q) &= w_M(q)\,\mathrm{d}q \\ &\text{-----------} \\ \mathrm{d}W_a(a) &= w_M(a)\,\mathrm{d}a \end{aligned} \tag{5'}$$

where $p, q, \ldots, a$ are possible complete sets of simultaneously measurable dynamical variables; dp, dq,...da, are the volume elements in the spaces $\mathcal{R}_p$, $\mathcal{R}(q)$ and $\mathcal{R}(a)$. The subscript M points to the fact that the distributions are related to the same macrosetting M.

It is obvious that the measurements leading to these distributions are performed on different specimens of the microsystem μ, μ', μ'',... since, according to the principle of complementarity it is impossible to detect in a quantum ensemble a specimen μ with simultaneously defined p and q. For this reason the above distributions are, generally speaking, incompatible with one another and are related to essentially different $p, q, \ldots a$ measurements.

This point demonstrates precisely the difference of principle with the classical ensemble in which there exists only one distribution (2) of the complete set of variables. On the other hand, we may notice that the distributions (5) in the quantum ensemble contain both the microsystem dynamical variables $p, q, \ldots a$ and the characteristic of the macrosetting marked by the subscript M. In this respect there is an entire analogy with the classical distribution (2).

All different quantum distributions (5′), belong to the same quantum ensemble. It is natural to suppose that there must exist a quantity that characterizes the quantum ensemble... irrespective of what kind of measurement is carried out on the particle or system μ. This quantity is the wave function Ψ_M which is defined by the macrosetting M. It is a function of the complete set of simultaneously measurable dynamical variables of the microsystem μ. Quantum mechanics is based on the assertion that all different distributions (5′) are completely determined by the wave function Ψ_M. If the wave function is given in the space $\mathscr{R}(p)$, then

$$w_M(p)=|\Psi_M(p)|^2 \tag{6}$$

and if it is given in the space $\mathscr{R}(q)$ then

$$w_M(q)=|\Psi_M(q)|^2. \tag{6′}$$

A more general version of the quantum ensemble arises when the macrosetting is insufficiently undefined so that it does not allow us to assign to this setting the wave function Ψ_M alone. There are possible several wave functions Ψ_{Mn}, $n=1, 2\ldots$, with a relative probability $\mathscr{P}_n(M)$. In this case the probability density is specified by means of the statistical operator introduced by von Neumann (also called the density matrix). Its matrix elements $R_M(q, q')$ are given by the formulė

$$R_M(q, q')=\sum_n \mathscr{P}_n(M)\, \Psi^*_{Mn}(q)\, \Psi_{Mn}(q') \tag{7}$$

where q and q' are the two values of the variables q of the space $\mathscr{R}(q)$. The diagonal term $(q=q')$ of this operator yields the probability density

$$w_M(q)=\sum_n \mathscr{P}_n(M)\,|\Psi_{Mn}(a)|^2. \tag{8}$$

The Neumann statistical operator $R_M(q, q')$ or the wave function $\Psi_M(q)$ provide a complete description of a quantum ensemble consisting of an infinite number of reproductions of the macrosetting M and the microsystem μ embedded in it. Of course, no matter whether this setting is created by an experimenter, as it occurs, for example, on accelerators, or this setting arises itself in nature. In other cases we may not be aware of the macroscopic conditions of M, and on the contrary, may judge of them from studying this quantum ensemble. An example is the investigation of cosmic rays when we attempt to establish the conditions of their origin from the analysis of their spectrum and composition.

Thus, according to the theory of quantum ensembles, the wave function $\Psi_M(q)$, or the statistical operator $R_M(q, q')$, are objective characteristics of such an ensemble and make it possible to predict the results of all possible measurements on the microsystem μ belonging to this ensemble.

In this connection it should be stressed that neither the wave function nor the statistical operator can be measured by measurements on a single specimen of the microsystem μ.

However, in principle, they can be determined from quantum ensemble measurements.

We illustrate this by an example. Let the wave function Ψ be a superposition of the two states Ψ_1, and Ψ_2 with definite values p_1 and p_2 of the momentum, respectively. In this case $\Psi=a_1\Psi_1+a_2\Psi_2$. Suppose that, performing measurements on the particle we obtain $p=p_1$. However, we would also obtain the same result in the case when the ensemble is characterized by the function $\varphi=\Psi_1$. Therefore, on the basis of this measurement we can not conclude whether the particle μ is described by the wave function Ψ or φ. Only by repeating measurements many times can we find out that the ensemble to which our particle belongs includes also the values $p=p_2$, and that there is an interference between these states. It is seen from this simple example that the wave function cannot belong to an individual particle in just the same way as the chance of a win is not the characteristic of the ticket itself but only shows that it belongs to the lottery as a whole.

4. Problem of Measurements

The problem of measurements and the related problem of the role of the observer were repeatedly investigated and discussed.

The formal aspect of the matter is well known. If the state of an ensemble is given by the wave function Ψ_M and the dynamical quantity under measurement is presented by the operator $\mathscr{L}$ with the eigenvalues L_n and the eigenfunctions Ψ_n (we consider the case when the system has a discrete spectrum and one degree of freedom), then Ψ_M can be represented as a linear superposition

$$\Psi_M = \sum_m C_m(M)\, \Psi_m. \tag{9}$$

If, after measurements, it turns out that $L = L_n$ this superposition is contracted to a new function

$$\Psi_M \to \Psi_n \tag{10}$$

As a result of reproductions of such measurements, there arises an ensemble described by the statistical operator

$$R(q, q') = \sum_m \mathscr{P}_m \Psi_m(q)\, \Psi_m(q). \tag{11}$$

In this case

$$\mathscr{P}_m = |C_m(M)|^2$$

is the probability of that the measurements in the initial ensemble will yield for L the value L_m.

There arises the question about the physical meaning of the contraction (10). Usually this contraction is thought of as an expression of the fact that after measurements of the quantity L on the microsystem μ the observer's information has changed. Such an interpretation is allowable when we deal with the experiment performed by the observer itself. However we suppose that the quantum description of phenomena remains valid as well for phenomena occurring without observer participation, e.g. radioactive decay of atoms in nature. With such a statement of the problem we are unable to refer to the change of the information. We are thus concerned with an objectively occurring phenomenon that may serve as the beginning of the chain of events up to the development of phenomena of macroscopic scale. The arising paradox can be resolved only if we include in the quantum description together with the microsystem the measuring instrument and look after its action. It is common opinion that actually this cannot be done because by uniting the instrument Π and the microsystem μ into one interacting quantum system $\mu+\Pi$ we should make use of a new instrument Π' by means of which we shall investigate the state of the system $\mu+\Pi$ etc.

The inconsistency of this conclusion will be seen from further analysis of the measurement problem. The measuring instrument consists of two parts of different purpose: an analyser $\mathscr{A}$ and a detector $\mathscr{D}$. By means of the analyser the initial ensemble described by the wave function Ψ_M (for simplicity we consider the case of a 'pure' ensemble) is decomposed into beams in such a way that the partial waves Ψ_m in the superposition (9) are separated and each wave transforms into a beam going to the corresponding detector placed in front of it. Figure 1 illustrates these considerations.

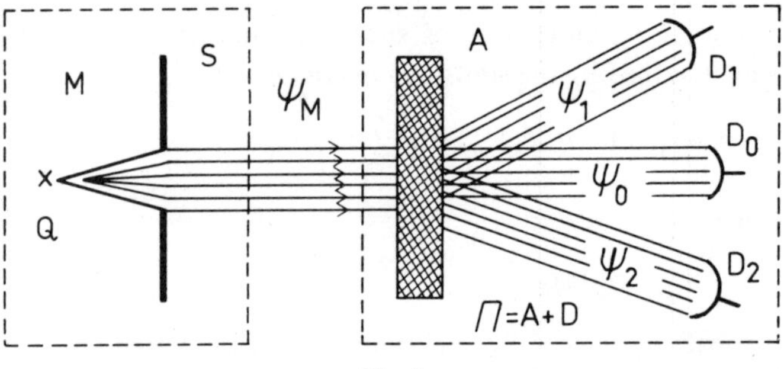

Fig. 1.

Q is a source of particles, s is a diaphragm with aperture, the left-hand area surrounded with the dotted line contains a macrosetting M that defines the ensemble of particles. The beam of these particles strikes a crystal $\mathscr{A}$ that yields a spectrum $\psi_1, \psi_2, \ldots$. This crystal serves as an analyser $\mathscr{A}$ and together with the detector $\mathscr{D}$ consisting of a few counters, $\mathscr{D}_1\mathscr{D}_2$ constitutes a measuring instrument $\Pi = \mathscr{A} + \mathscr{D}$. This instrument is shown in the right-hand part of the figure.

It is essential that the detector is a macroscopically unstable system. This is clear from the argument that the microparticle μ may have neither the energy nor the momentum sufficient for a macroscopic device to be put into operation (for example, the microparticle cannot deflect the pointer in a meter). If the microparticle had such an ability then we would consider it as a macroscopic object. However, a microparticle can induce a macroscopic phenomenon if the device with which it interacts is macroscopically unstable. In this case the interaction of the particle μ with the detector begins, of course, at the level of a microscopic quantum phenomenon that then develops up to the level of a macroscopic phenomenon. This event is precisely the contraction of the wave packet that occurs due to the result $L=L_m$ obtained from measurement. The 'contraction' (10) considered from the microscopic point of view has the character of an explosion, initiated by an atomic quantum phenomenon. Notice that all detectors are really macroscopic unstable systems (electrically, chemically or thermodynamically). It is enough to recall a Geiger counter, the blackening of grains in a photographic emulsion, a Wilson cloud chamber or a bubble chamber. Summarizing, we may say that the quantum measurement ('contraction') consists in initiating a macroscopic process in a macroscopically unstable system. In this analysis not one word has been said about the observer. It is clear that the situation described consists in that the macrosetting $\mathfrak{M}=M+\mathscr{D}$ which contains a macroscopically unstable element $\mathscr{D}$ can be realized in nature in itself without any participation of the experimenter. Supplementing of the instrument $\Pi=\mathscr{A}+\mathscr{D}$ with a new classical one Π' will, of course, introduce nothing new since in $\mathscr{D}$ there already developed a macroscopic event which demands no quantum description. From this viewpoint the operation of the measuring instrument Π can be considered by the methods of quantum mechanics. However, uniting the microparticle μ and the measuring instrument we are obliged to describe the new ensemble by means of the statistical operator $R(q, Q, | q'Q')$ depending on the dynamical variables of the particle q and the variables Q describing the measuring instrument, more exactly, the detector $\mathscr{D}$.

The wave function alone will, in this case, be insufficient since the detector is a macroscopic system. At the initial moment $t=0$ the statistical operator R can be represented in the form of the product

$$R_0(q, Q, q'Q')=r_0(q, q')\, R_0(Q, Q'), \tag{13}$$

where r_0 is the statistical operator describing the microsystem at the moment $t=0$ and R_0 is the operator describing the detector also at $t=0$.

Owing to the effect of the microsystem on the detector, with the passage of time, the statistical operator R will change. Knowing at the initial moment its matrix elements (13) from the equation

$$\frac{\partial \hat{R}}{\partial t}+[\hat{H}\hat{R}]=0 \tag{14}$$

we can find the statistical operator R at $t\to\infty$. In Equation (14) the operator $\hat{H}$ denotes the interaction energy between the particle μ and the detector $\mathcal{D}$, [ ] are the Poisson brackets. If the detector $\mathcal{D}$ is unstable then at $t\to\infty$ the matrix $R(q, Q, q', Q', t)$ will describe a new macroscopic phenomenon, e.g. an electric discharge in the Geiger counter.

In conclusion we give a simple example illustrating the ideas presented in this section.

Let the state of the microparticle μ in an ensemble be described by a standing wave:

$$\Psi_M(x)=\frac{1}{\sqrt{2}}(e^{ikx}+e^{-ikx}), \tag{15}$$

where x is the particle coordinate. We consider a detector which should answer the question about the sign of the particle momentum k. As such

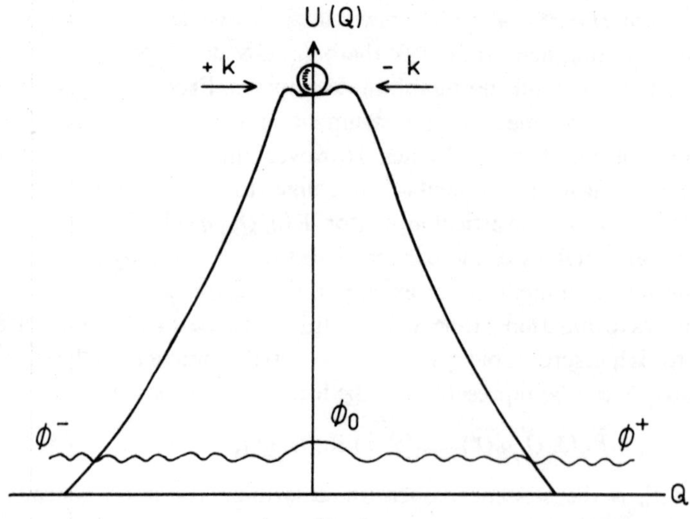

Fig. 2.

a detector we take a macroscopic ball of mass M and place it at the top of a cone. This ball is supposed to scatter μ particles. Since the ball is in unstable equilibrium the scattering of a particle with momentum k will push if off from the top of the cone, say, to the right, and the scattering of a particle with momentum to the left. The particle scattering initiating the fall of the ball is a quantum phenomenon. However, when falling the ball acquires a large energy and the quantum phenomenon develops up to the level of a macroscopic phenomenon. This is precisely the action of our detector. Figure 2 is an illustration of the above considerations, where $V(a)$ is the potential energy of the ball, Q is its coordinate. The scattering of a particle with momentum $\pm k$ gives an impact to the ball and knocks it out from within a small hole at the top of the cone. The stage of the ball far off the cone (at $|Q| \to \infty$) is, within an arbitrary accuracy, described by the calssical action function $S(Q)$. ϕ^+ and ϕ^- are the asymptotic wave functions of the ball equal to $\exp \pm (i/\hbar)\, S(Q)$. They do not interfere with each other.

Another example, which we can calculate, may be a thermodynamically unstable system. Let us assume that in the sites of a plane lattice there are located charged particles that oscillate about the equilibrium state in the plane x, y. We also assume that the initial state for $t=0$ is such that the oscillations along the axis ox occur at a certain temperature θ and the oscillations along the axis oy are frozen in such a way that their tem-

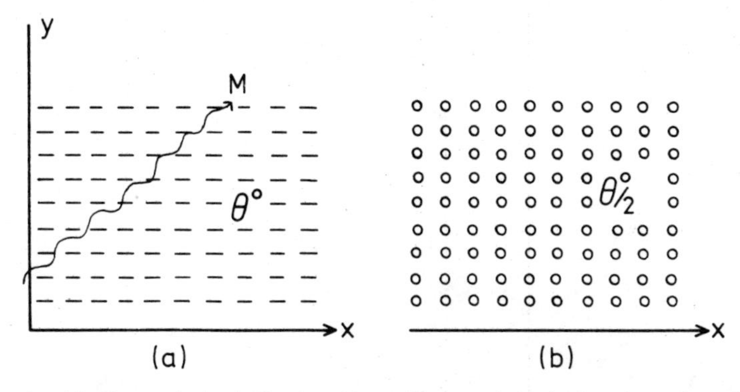

Fig. 3. (a) Thermodynamically unstable oscillation of the lattice along the axis ox at a temperature θ while the oscillation along the axis oy occurs at $\theta=0$. (b) Circular oscillation of the lattice after the interaction with the μ particle, the temperature is now $\theta/2$.

perature is zero, $\theta'=0$. Thus, the system is thermodynamically unstable.

If a microparticle μ having magnetic moment penetrates into this medium it serves as a dust particle that provides exchange of energy between two degrees of freedom of the lattice particles, and at $t\to\infty$ the temperature of the lattice falls down to $\theta/2$. The action of the detector consists, in this case, in lowering the temperature of the lattice.

Detailed mathematical calculations for both kinds of detector, mechanically and thermodynamically unstable ones, are given in the papers of the author [2, 4].

BIBLIOGRAPHY

[1] Blokhintsev, D. I., *Mécanique quantique*. Edit. Masson et Cie, Paris 1967; *Quantum Mechanics*, D. Reidel Publ. Company, Dordrecht, 1964.

[2] Blokhintsev, D. I., *Principes essentiels de la Mécanique quantique*. Monographies Dunod, Paris (1967); *The Philosophy of Quantum Mechanics*, D. Reidel Publ. Company, Dordrecht, 1968.

[3] Blokhintsev, D. I., 'Otwet academiku V. A. Foku', *Voprosy filosofii* N6 (1952), 171.

[4] Blokhintsev, D. I., *Uspekhi fizicheskih nauk* **95** (1958), 75.

SATOSI WATANABÉ

CONDITIONAL PROBABILITY IN WAVE MECHANICS*

In his paper of 1948 [1] and in his book of 1956 [2] Louis de Broglie stressed that the ordinary relationship between joint probability and conditional probability no longer holds in wave mechanics, and that the difference lies above all in the definition of the joint probability of two physical quantities which are not simultaneously measurable. What is still more important, he showed that the double solution theory is not subject to this difficulty. He is certainly right in emphasizing this problem of joint probability and conditional probability, since it is the centre of the curious features of wave mechanics, which should be dealt with first if we wish to attempt an improvement of its ordinary interpretation.

We do not propose, however, to introduce any theoretical novelty with regard to probability in this paper. We will remain within the framework of ordinary wave mechanics and will analyze not only the theoretical positions of conditional probability and joint probability, but also the algebraic structure, that is the 'logic' which exists in the set of experimental propositions of wave mechanics.

We start by the introduction of three types of 'products' which can be defined amongst the 'propositions', A, B, C, etc.

(i) A 'followed by' B:

$$B \cdot A \tag{1}$$

(ii) A 'and then' B:

$$B \overset{\leftarrow}{\cap} A \tag{2}$$

(iii) A 'and' B (conjunction):

$$B \cap A. \tag{3}$$

In general the product 'followed by' obeys the associative law, but does not necessarily obey the idempotential law nor the commutative law. If a proposition obeys the idempotential law, $A \cdot A = A$, we call it 'simple proposition'. If two propositions satisfy the commutative law, they are

M. Flato et al. (eds.), Quantum Mechanics, Determinism, Causality, and Particles, 159–165. *All Rights Reserved. Copyright © 1976 by D. Reidel Publishing Company, Dordrecht-Holland.*

said to be 'compatible'. The compatibility relationship is symmetrical and reflexive but not transitive.

The product 'and then' in general obeys neither the idempotential law nor the associative law nor the commutative law. If the proposition is simple, it obeys the idempotential law with regard to the product 'and then' also. If A and B are simple and compatible, they satisfy the commutative law in the sense of the product 'and then'. If three propositions are simple and compatible in pairs, they satisfy the associative law with regard to the product 'and then'.

The product 'and' or the conjunction satisfies the associative law. If the propositions intervening are simple (not necessarily compatible), all three laws hold good with regard to the product 'and'. If A and B are simple propositions, the product A 'and' B is also a simple proposition.

A physical quantity Q can in general be expressed in the form:

$$Q=\sum_k q_k P[\varphi_k], \tag{4}$$

where q_k is an eigenvalue of Q, and $P(\varphi_k]$ is the projection operator for the eigenfunction (vector) corresponding to q_k. If there is a multiplicity in the eigenvalues we can write

$$Q=\sum_\mu q_\mu P[M_\mu], \tag{5}$$

where $P[M_\mu]$ is the projection operator for the subspace determined by the eigenfunctions corresponding to the degenerate eigenvalue q_μ. The projection operator is a Hermitian matrix having eigenvalue 0 or 1, that is

$$P[M_\mu]^2=P[M_\mu]. \tag{6}$$

The dimension of the subspace is given by

$$D_\mu=\text{trace } P[M_\mu]. \tag{7}$$

The observation proposition is of the form: 'The quantity Q has the value q_μ'. We can therefore make the operator $P[M_\mu]$ correspond to this proposition. The propositions considered above are not limited to this type of observation propositions. But we always suppose that there is a matrix corresponding to a proposition, and we use the same symbols A, B, C, etc also for the matrices corresponding to the propositions.

In terms of matrices, the different connectives introduced above are written:

$$A \cdot B = AB \tag{8}$$

$$A \overleftarrow{\cap} B = ABA \tag{9}$$

$$A \cap B = \ldots ABABABA \ldots \tag{10}$$

If A and B are projection operators, the product 'followed by' is not necessarily a projection operator. If A is a projection operator, it satisfies the idempotential law in the sense of the product 'followed by' as a consequence of (6). The observation proposition is therefore a simple proposition. Compatibility corresponds to the commutative property of matrices. The product 'and then' of two Hermitian operators is Hermitian. But the product 'and then' of two projection operators is not necessarily a projection operator, but it is a Hermitian operator with nonnegative eigenvalues. All that has been said about the idempotential, commutative, and associative laws with regard to the product 'and then' follows from the definition of this connective in terms of matrices. The fact that the product 'and' satisfies all three laws, provided that the operators concerned are projection operators, requires a proof, but the reader will find this in my book *Knowing and Guessing* [3]. It is important to note that if we limit ourselves to commutative operators, all three connectives become identical.

It is to be noted that the class of Hermitian operators with nonnegative eigenvalues is closed with regard to the connective 'and then', and that the class of projection operators is closed with regard to the connective 'and'. The class of commutative projection operators is closed with regard to each of the three connectives' that are equivalent in this case.

The operator 'zero' $\emptyset$ and the operator 'all' $\square$ satisfy the following laws:

$$\begin{array}{ll} \emptyset \cdot A = A \cdot \emptyset = \emptyset & \square \cdot A = A \cdot \square = A \\ \emptyset \overleftarrow{\cap} A = A \overleftarrow{\cap} \emptyset = \emptyset & \square \overleftarrow{\cap} A = A \overleftarrow{\cap} \square = A \\ \emptyset \cap A = A \cap \emptyset = \emptyset & \square \cap A = A \cap \square = A. \end{array} \tag{11}$$

The matrix corresponding to $\emptyset$ is the matrix with all coefficients zero, and the matrix corresponding to $\square$ is the identity matrix I. The matrices $\emptyset$ and $\square$ are Hermitian, simple, and commute with all matrices.

The negation operation is defined by the following laws:

(i) $\neg(\neg A) = A$
(ii) $A \cap B = A$ is equivalent to $\neg A \cap \neg B = \neg B$ (12)
(iii) If $\neg A \cap A = A$, then $A = \emptyset$.

In terms of matrices, the negation can be written:

$$A = I - A. \tag{13}$$

The connective 'or' or disjunction can be defined by:

$$A \cup B = \neg(\neg A \cap \neg B). \tag{14}$$

It can be shown that the simple propositions obey the absorptive law:

$$A \cap (A \cup B) = A. \tag{15}$$

We have in addition formulae which are obtained from those already given, by simultaneous replacement of $\cap$ by $\cup$, $\cup$ by $\cap$, and a proposition by its negation. We can thus show that the simple propositions form a lattice. We cannot, however, prove the distributative law. The lattice is therefore nondistributive. If we restrict the discussion to those propositions which are mutually commutative in pairs, the lattice becomes distributive.

The probability that we shall obtain the affirmative result by observation of A in a system which is in state Z is given by:

$$p(A; Z) = \text{trace}(A \cdot Z), \tag{16}$$

where Z is a Hermitian operator with nonnegative eigenvalues which is normalized:

$$\text{trace } Z = 1. \tag{17}$$

A special state Z_0 known as the 'neutral state' is defined by:

$$Z_0 = I/\text{trace } I. \tag{18}$$

If we consider only simple operators and the neutral state, it can be shown that:

$$p(A; Z_0) + p(B; Z_0) = p(A \cap B; Z_0) + p(A \cup B; Z_0) \tag{19}$$

satisfying the fundamental axiom of probability:

$$\text{prob}(A)+\text{prob}(B)=\text{prob}(A\cap B)+\text{prob}(A\cup B) \tag{20}$$

This is evident since $p(A; Z_0)$ is proportional to the dimension of the subspace corresponding to A.

As we have shown elsewhere (*Knowing and Guessing*, pp. 367, 464 and 504) there is however an anomaly even in the case of the neutral state. If we derive four conditional probabilities:

$$\begin{aligned} p(A \mid C; Z_0) &= p(A\cap C; Z_0)/p(C; Z_0) \\ p(B \mid C; Z_0) &= p(B\cap C; Z_0)/p(C; Z_0) \\ p(A\cap B \mid C; Z_0) &= p(A\cap B\cap C; Z_0)/p(C; Z_0) \\ p(A\cup B \mid C; Z_0) &= p((A\cap B)\cap C; Z_0)/p(C; Z_0) \end{aligned} \tag{21}$$

they do not satisfy the fundamental axiom of probability (20) unless the propositions A, B, and C satisfy the distributative law. Since we do not in general have this law included in the observation propositions, we conclude that conditional probabilities do not exist even in the most favourable case of the neutral state Z_0. If we restrict the discussion to simple compatible propositions, the quantities defined in (21) behave as true conditional probabilities.

It must be noted that use can be made of the fact (*Knowing and Guessing*, pp. 450 and 494) that it is possible to define a (nonconditional) probability by means of (19) in order to show that in the finite dimensional case the lattice is modular, i.e. that:

$$A\cap(B\cup C)=(A\cap B)\cup C \tag{22}$$

if $C=C\cap A$.

If we use a state Z which is not neutral in (16), we cannot prove a formula of type (19), showing that probability cannot in general be defined in this lattice. If we restrict the discussion to distributative sublattice, we can define not only the simple probability but also the conditional probability.

We come finally to the question of successive observations. For this purpose it is necessary to know how the state Z changes after an observation. If we make an observation to determine whether A or not-A on a system in state Z, and if we obtain an affirmative result, Z becomes Z' given by:

$$Z' = AZA/\text{trace}(AZA). \tag{23}$$

The probability of obtaining this result is, from (16):

$$p(A; Z) = \text{trace}\,(AZ) = \text{trace}\,(AZA). \tag{24}$$

We now make a second observation to determine whether B or not-B on the system which is now in state Z'; the probability of an affirmative result will be:

$$\begin{aligned} p(B; Z') &= \text{trace}\,(BZ') = \text{trace}\,(BAZA)/\text{trace}\,(AZA) \\ &= \text{trace}\,(ABAZ)/\text{trace}\,(AZA). \end{aligned} \tag{25}$$

The joint probability of result A and then result B will be given by the product of the two probabilities (24) and (25), i.e.

$$p(A \text{ and then } B; Z) = \text{trace}\,(ABA \cdot Z) \tag{26}$$

This joint probability is thus formally equivalent to the probability of observation of ABA on the initial state Z (although ABA is not a projection operator). It will be seen why we have called ABA 'A and then B'. It is precisely this joint probability that L. de Broglie considered in his discussion. In wave mechanics, according to the usual interpretation, we can construct three joint probabilities for A and B, all different: the probability of A and then B: trace $(ABA \cdot Z)$; the probability of B and then A: trace $(BAB\,Z)$; and the probability of A and B: trace $(\ldots ABABAB \ldots Z)$. These three coincide when A and B are compatible.

We finish this short paper by mentioning that Z is the sum of several projection operators with nonnegative weights. Thus, for a discussion of the logical structure of the propositions, there is no need to distinguish Z from other operators such as A, B, C, etc. The 'logic' is thus isomorphic with the algebra of projection operators and the probability is given by the trace of a product of projection operators. We would also mention that Z' becomes the projection operator A if Z is Z_0, except for normalization.

University of Hawaii

NOTE

* Translated from the French by N. Corcoran.

BIBLIOGRAPHY

[1] de Broglie, Louis, *Revue Scientifique* (1948), 259.
[2] de Broglie, Louis, *Une tentative d'interprétation causale et non-linéaire de la mécanique ondulatoire*, Gauthier-Villars, Paris, 1956.
[3] Watanabé, Satosi, *Knowing and Guessing*, Wiley, New York, 1969.

ROBERT L. ANDERSON AND JOHN G. NAGEL

CLASSICAL ELECTRODYNAMICS IN TERMS OF A DIRECT INTERPARTICLE HAMILTONIAN

ABSTRACT. A usual-time Hamiltonian formalism for the Wheeler-Feynman Theory of Classical Electrodynamics is presented in which: (i) momenta canonically conjugate to the particle coordinates are defined and the associated problem of the inversion of these defining relations for particle velocities in terms of the particle coordinates and canonical momenta is formally solved; (ii) the equations of motion for the canonical variables are expressed in the Hamiltonian form; (iii) the usual definition and properties of the Poisson bracket are preserved. There are two new and essential properties of the Hamiltonian in this theory: (i) it is a functional of the canonical variables, as well as a function of them; (ii) in general, it is not conserved in time.

1. INTRODUCTION

The physical importance of the Wheeler-Feynman direct interparticle action theory of classical electrodynamics[1-4] lies in the fact that: (i) the dynamical equations of motion which characterize the theory can be derived from the relativistically invariant Fokker action principle[3]; (ii) all the results of the Maxwell-Lorentz electrodynamics can be reproduced by employing the Wheeler-Feynman absorber theory of radiation[4]; (iii) the solutions of Dirac[5] and Gupta[6] to the problem of the elimination of the self-energy are built into this theory and the physical origins of these solutions are described by the absorber theory of radiation.

In the past, the obstacle in this theory to the development of a usual-time Hamiltonian formalism in which particle coordinates appear as canonical variables was the proper identification of observable momenta canonically conjugate to the particle coordinates[7]. The essential mathematical problem is to invert the defining relations for the canonical momenta and express the particle velocities in terms of the canonical variables. The essential physical problem is to argue the observability of the canonical momenta[8]. In this paper a formal solution to this problem is presented which then permits a usual-time Hamiltonian formulation of the theory.

M. Flato et al. (eds.), Quantum Mechanics, Determinism, Causality, and Particles, 167–178. *All Rights Reserved. Copyright © 1976 by D. Reidel Publishing Company, Dordrecht-Holland.*

2. ONE-PARTICLE LAGRANGIANS, MOMENTA, AND HAMILTONIANS

First, it is necessary to state the Wheeler-Feynman theory, hereafter referred to as the W-F theory, in the language of one-particle Lagrangians, momenta, and Hamiltonians. The W-F theory[4] is characterized by a Poincaré invariant n-particle direct interparticle functional $\hat{L}(t_1, t_2)$ of the form[9]

$$\hat{L}(t_1, t_2) = \hat{L}\left[x_a\begin{pmatrix} t_2 \\ t \\ t_1 \end{pmatrix}, \begin{matrix} n \\ a \\ 1 \end{matrix}\right] =$$

$$= \sum_{a=1}^{n} \int_{t_1}^{t_2} \mathrm{d}t \, \mathscr{L}_a(x'_a(t)) +$$

$$+\tfrac{1}{2} \sum_{a,b=1}^{n}{}' \int_{t_1}^{t_2} \mathrm{d}t \int_{t_1}^{t_2} \mathrm{d}t' \, \mathscr{L}_{ab}(\xi_{ab}(t, t'), \xi'_{ab}(t, t')), \tag{1}$$

where

$$\mathscr{L}_a(t) = g_{00} m_a c \sqrt{g_{00} x'_{a\mu}(t) \, x'^{\mu}_a(t)}, \tag{2}$$

$$\mathscr{L}_{ab}(t, t') = \frac{e_a e_b}{c} \, x'_{a\mu}(t) \, \delta(\xi^2_{ab}(t, t')) \, x'^{\mu}_b(t'), \tag{3}$$

$$\xi_{ab}(t, t') = x_a(t) - x_b(t') \tag{4}$$

and the parameter t is equal to the usual time. Because the interaction is not propagated instantaneously from one particle to another the direct interparticle functional $\hat{L}(t_1, t_2)$ plays a dual role[10]. First, the requirement that in the limit $t_1 \to -\infty$, $t_2 \to +\infty$ the vanishing of the fixed end-point and interval of integration variation of $\hat{L}(t_1, t_2)$ yields the equations of motion (Fokker action principle). This requirement can be stated as

$$\lim_{\substack{t_2 \to +\infty \\ t_1 \to -\infty}} \frac{\delta \hat{L}}{\delta x^{\mu}_a} = 0, \tag{5}$$

which implies the 4-vector equations of motion

$$\frac{\partial L_a(t)}{\partial x^{\mu}_a(t)} - \frac{\mathrm{d}}{\mathrm{d}t}\left(\frac{\partial L_a(t)}{\partial x'^{\mu}_a(t)}\right) = 0, \qquad (a = 1, \ldots, n) \tag{6}$$

where

$$L_a(t)=\mathscr{L}_a(t)+\sum_{b\neq a}\int_{-\infty}^{\infty} \mathrm{d}t'\, \mathscr{L}_{ab}(t,t'). \tag{7}$$

Formally, $L_a(t)$ may then be considered as a one-particle Lagrangian [11]. If one-particle 4-momenta $p_{a\mu}(t)$ are defined by the relation

$$p_{a\mu}(t)=\frac{\partial L_a(t)}{\partial x_a'^{\mu}(t)}=\frac{m_a c x'_{a\mu}(t)}{\sqrt{g_{00}x_a'^{\nu}(t)\, x'_{a\nu}(t)}}+\frac{e_a}{c}A_{a\mu}(t), \tag{8}$$

where

$$A_{a\mu}(t)=\sum_{b\neq a} e_b\int_{-\infty}^{\infty} \mathrm{d}t'\, \delta(\xi_{ab}^2(t,t'))\, x_b'^{\mu}(t'), \tag{9}$$

then formally there exists a one-particle Hamiltonian $H_a(t)$ defined by

$$H_a(t)=p_a^0(t)\, c \tag{10}$$

and a set of Hamilton's equation for each particle, i.e.,

$$\frac{\partial H_a(t)}{\partial x_a^i(t)}=-p'_{ai}(t),\qquad \frac{\partial H_a(t)}{\partial p_a^i(t)}=x'_{ai}(t). \tag{11}$$

This follows immediately by comparing Equations (1)–(11) with the corresponding equations for the theory of a charged particle in a given external field [12].

Second, the fact that the general variation of $\hat{L}(t_1,t_2)$ vanishes identically under an infinitesimal Poincaré transformation coupled with its evaluation along physically allowed particle trajectories yields the ten conservation laws [13] of 4-momentum, angular momentum and center of energy [14]. This furnishes, in particular, the connection between the one-particle momenta and the conserved 4-momentum, i.e.,

$$p^{\mu}(t)=\sum_a p_a^{\mu}(t)+\sum_{a,b}{}'\frac{e_a e_b}{c}\left\{\int_t^{\infty}\mathrm{d}t'\int_{-\infty}^{t}\mathrm{d}t-\int_{-\infty}^{t}\mathrm{d}t\int_t^{\infty}\mathrm{d}t'\right\}\times \tag{12}$$
$$\times\, x_a'^{\nu}(t)\,\delta'(\xi_{ab}^2(t,t'))\,\xi_{ab}^{\mu}(t,t')\,x'_{b\nu}(t').$$

This expression shows that, in general, the one-particle momenta are neither separately nor when summed conserved in time.

3. Usual-time Hamiltonian Formalism

Now, we are in a position to state the essential physical result of this paper, namely, our assertation that the usual-time Hamiltonian $H(t)$ for the W-F theory is given by the following equation

$$H(t)=\sum_a H_a(t). \tag{13}$$

The validity of this assertion rests on three properties of $H(t)$. First, there exists a set of canonical momenta in the sense of a Hamiltonian formalism, i.e., the defining relations $p_{ai}=\partial L_a/\partial x_a'^i$ for the set of one-particle 3-momenta are invertible for the $x_a'^i$'s, each of which becomes a function of the corresponding x_a^j's and p_a^j's and a functional of all the x_b^j's and p_b^j's. This is the essential mathematical result of this paper and is proven in subsequent paragraphs. This result implies that $H(t)$ can be expressed as a function and a functional of the set of canonical variables $\{x_a^i, p_a^i\}_{a=1}^n$. Second, $H(t)$ reproduces the equations of motion in the Hamiltonian form. This follows because $H(t)$ is a function of any canonical variable only through its corresponding one-particle Hamiltonian. Otherwise, $H(t)$ is a functional of each canonical variable, but this functional dependence does not make any contribution under the operation of partial differentiation w.r.t. any canonical variable. As previously pointed out, this property follows from the mathematics of a charged particle in a given external field. Third, the time rate of change of any observable F is given by

$$\frac{\mathrm{d}F(t)}{\mathrm{d}t}=\{F(t), H(t)\}+\frac{\partial F(t)}{\partial t}, \tag{14}$$

where

$$\{\cdot,\cdot\}\equiv -g_{00}\left(\frac{\partial\cdot}{\partial x_a^i}\frac{\partial\cdot}{\partial p_{ai}}-\frac{\partial\cdot}{\partial p_{ai}}\frac{\partial\cdot}{\partial x_a^i}\right) \tag{15}$$

is the usual Poisson bracket. This follows immediately from Hamilton's equations of motion.

4. Invertibility of the Defining Relations for the Canonical Momenta

We shall now establish the invertibility of the defining relations for the one-particle 3-momenta $p_{ai}=\partial L_a/\partial x_a'^i$ and hence show they are the momenta canonically conjugate to the particle space coordinates x_a^i's. In order to invert these defining relations, we first pass to the multiple-proper time parameterization of the system. Rewriting Equation (8) in this parameterization yields

$$\Pi_a^\mu(\tau_a) = m_a \dot{y}_a^\mu(\tau_a) + \\ + \sum_{b \neq a} \frac{e_a e_b}{c} \int_{-\infty}^{\infty} \mathrm{d}\tau_b \, \delta([y_a(\tau_a) - y_b(\tau_b)]^2) \, \dot{y}_b^\mu(\tau_b) \tag{16}$$

where

$$\Pi_a^\mu(\tau_a) \equiv p_a^\mu(t), \quad y_a^\mu(\tau_a) \equiv x_a^\mu(t), \quad \mathrm{d}\tau_a = \frac{1}{c}\sqrt{g_{00} x_a'^\mu(t)\, x_{a\mu}'(t)}\, \mathrm{d}t, \tag{17}$$

and

$$\cdot = \frac{\mathrm{d}}{\mathrm{d}\tau_a}.$$

Equation (16) can now be written as

$$\frac{\Pi_a^\mu(\tau_a)}{m_a c} = (1 + \mathrm{I}) \frac{\dot{y}_a^\mu(\tau_a)}{c}, \tag{19}$$

where we have introduced the integral operator

$$\mathrm{I} = \sum_{b \neq \cdot} \frac{e_o e_b}{m_o c} \int_{-\infty}^{\infty} \mathrm{d}\tau_b \, \delta([y_0(\tau_0) - y_b(\tau_b)]^2), \tag{20}$$

such that its action is defined as follows

$$\mathrm{If}\,(m_a, y_a(\tau_a), \dot{y}_a(\tau_a), \Pi_a(\tau_a)) = \\ = \sum_{b \neq a} \frac{e_a e_b}{m_a c} \int_{-\infty}^{\infty} \mathrm{d}\tau_a \, \delta([y_a(\tau_a) - y_b(\tau_b)]^2) \times \\ \times f(m_b, y_b(\tau_b)\, \dot{y}_b(\tau_b), \Pi_b(\tau_b)). \tag{21}$$

Now, formally if $(1 + \mathrm{I})^{-1}$ exists, we have

$$\frac{\dot{y}_a^\mu}{c} = (1 + \mathrm{I})^{-1} \frac{\Pi_a^\mu}{m_a c} \tag{22}$$

which implies

$$\frac{x_a'^{\mu}}{c} = \frac{(1+\mathrm{I})^{-1}\,\Pi_a^{\mu}/m_a c}{\sqrt{1+[(1+\mathrm{I})^{-1}\,\boldsymbol{\Pi}_a/m_a c]^2}}, \tag{23}$$

where

$$\boldsymbol{\Pi}_a = (\Pi_a^1, \Pi_a^2, \Pi_a^3).$$

This is applicable for strong as well as weak coupling if $(1+\mathrm{I})^{-1}$ exists. Now, in order to obtain an equation which determines the action of $(1+\mathrm{I})^{-1}$ on the canonical variables $\{x_a^i, p_a^i\}_{a=1}^n$ in terms of these variables, we first rewrite the integral operator I in a form which involves the parameter t and the $\mathbf{x}_a(t)$'s and $\mathbf{x}_a'(t)$'s, namely,

$$\mathrm{I} = \sum_{b\neq\cdot} \int_{-\infty}^{\infty} \mathrm{d}t' \frac{e_a e_b \frac{1}{2}[\delta(t'-t_A') + \delta(t'-t_R')]\sqrt{1-(\mathbf{x}_b(t')/c)^2}}{m_o c^2 |\boldsymbol{\xi}_{\cdot b}(t,t')|\,[1+[\mathrm{sgn}(t'-t)\,\hat{\boldsymbol{\xi}}_{\cdot b}(t,t')\cdot \mathbf{x}_b'(t')/c]}, \tag{25}$$

where

$$\hat{\boldsymbol{\xi}}_{\cdot b}(t,t') \equiv \frac{\boldsymbol{\xi}_{\cdot b}(t,t')}{|\boldsymbol{\xi}_{\cdot b}(t,t')|}, \tag{26}$$

and

$$t'_{\substack{A\\R}} \equiv t \pm \frac{1}{c}\,|\boldsymbol{\xi}_{\cdot b}(t,t')|. \tag{27}$$

Note the denominator in Equation (25) never vanishes because all particle world-lines are time-like. Equations (23) and (25) now combine to yield an integral equation for $(1+\mathrm{I})^{-1}$ in terms of the x_a^i's and p_a^i's, namely,

$$1 = (1+\mathrm{I})^{-1}\left\{1 + \sum_{b\neq\cdot}\int_{-\infty}^{\infty} \mathrm{d}t' \frac{e_o e_b \frac{1}{2}[\delta(t'-t_A') + \delta(t'-t_R')]}{m_o c^{\tau} |\boldsymbol{\xi}_{\cdot b}(t,t')|} \times \right.$$

$$\left. \times \frac{\omega_b(t')}{1+[\mathrm{sgn}(t'-t)]\,\hat{\boldsymbol{\xi}}_{\cdot b}(t,t')\cdot[(1+\mathrm{I})^{-1}\,\mathbf{p}_b(t')/m_b c]\,\omega_b(t')}\right\}, \tag{28}$$

where

$$\omega_b(t') = (1+[(1+\mathrm{I})^{-1}\,\mathbf{p}_b(t')/m_b c]^2)^{-1/2}. \tag{29}$$

Under the assumption that $(1+\mathrm{I})^{-1}$ exists, the integral equation (28) is exact and hence the solution for $(1+\mathrm{I})^{-1}$ inserted into Equation (23) yields the desired inversion of the equations $p_{ai}(t)=\partial L_a(t)/\partial x_a'^i(t)$, irrespective of the strength of the interaction[15].

If

$$|\mathrm{If}\,(m_a,\, \chi_a(t),\, p_a(t);\, t)|<|f\,(m_a,\, \chi_a(t),\, p_a(t);\, t)| \tag{30}$$

for all t, then Equation (28) can be rewritten as

$$\begin{aligned}\mathrm{I}=\sum_{b\neq\cdot}\int_{-\infty}^{\infty}\mathrm{d}t'\,&\frac{e_o e_b \frac{1}{2}[\delta(t'-t_A')+\delta(t'-t_R')]}{m_o c^2\,|\boldsymbol{\xi}_{\cdot b}(t,t')|}\times\\ &\times\omega_b(t')\sum_{s=0}^{\infty}(-1)^s\Bigg([\mathrm{sgn}(t'-t)]\,\boldsymbol{\xi}_{\cdot b}(t,t')\cdot\\ &\cdot\left(\frac{1}{1+\mathrm{I}}\frac{\mathbf{p}_b(t')}{m_b c}\right)\omega_b(t')\Bigg)^s,\end{aligned} \tag{31}$$

where

$$\begin{aligned}\omega_b(t')=&\left(\frac{m_b^2c^2+\mathbf{p}_b(t')^2}{m_b^2c^2}\right)^{-1/2}\sum_{k=0}^{\infty}\binom{-\frac{1}{2}}{k}\times\\ &\times\left[\frac{\dfrac{-2p_{bi}(t')}{m_b c}\dfrac{\mathrm{I}}{1+\mathrm{I}}\dfrac{p_b^i(t')}{m_b c}+\left(\dfrac{\mathrm{I}}{1+\mathrm{I}}\dfrac{\mathbf{p}_b(t')}{m_b c}\right)^2}{\dfrac{m_b^2c^2+\mathbf{p}_b^2(t')}{m_b^2c^2}}\right]^k\end{aligned} \tag{32}$$

and

$$(1+\mathrm{I})^{-1}=\frac{1}{1+\mathrm{I}}=\sum_{r=0}^{\infty}(-1)^r\,\mathrm{I}^r. \tag{33}$$

Equation (31) evidently holds for many relativistic particle configurations but the rigorous limits of validity of it must yet be established. We are presently investigating, using both analysis and computer experiments, the connection between Equation (31) and the particle configuration in which the set of dimensionless dynamical quantities

$$\eta_{ab}^{\pm}\equiv\frac{e_a e_b\sqrt{1-(\mathbf{x}_b'/c)^2}}{m_a c^2|\boldsymbol{\xi}_{ab}|\,[1\pm\boldsymbol{\xi}_{ab}\cdot\mathbf{x}_b'/c]}, \tag{34}$$

is such that $\eta_{ab} < 1$ for almost all velocities and advanced and retarded separations. This condition is of interest because it would insure that Equation (30) would apply for almost all times.

In the case that Equation (31) holds, its solution[16] when substituted into Equation (23), then constitutes the solution to the problem of the inversion of the defining relations for the canonical momenta for the W-F theory. The expression for the Hamiltonian $H(t)$ in terms of the canonical variables and t readily follows.

5. Observability of the Classical Canonical Momenta

The observability of the canonical momenta $\mathbf{p}_a(t)$ reduces to that of the one-particle electromagnetic potentials $\mathbf{A}_a(t)$. These potentials do not involve, as does the conserved momentum, the electromagnetic momenta in transit between the particles at the time t, and are therefore, in principle, observable. In the last section we comment on the situation that pertains in any quantum version of this theory.

6. Two-Hamiltonian Theory

For interacting systems in the W-F theory, we then have in effect two Hamiltonians. One is the quantity $\sum_a H_a(t)$ which plays the role of generating the canonical equations of motion and is in general not conserved while the second $p^\circ c$ generates the one-parameter Poincaré time-translation group and is conserved. This situation translates into the statement that Hamilton's equations of motion for this theory are not actively Poincaré covariant[17].

7. Remarks on the Problem of the Quantization of the W-F Theory

The intertwining of Einstein's[18] Principle of Special Relativity and the usual-time Hamiltonian formalism of classical particle mechanics presented here suggests a starting point from which to attempt to build a relativistic quantum theory of directly interacting charged particles either in the tradition of the wavemechanics of Einstein[19], de Broglie[20], and Schrödinger[21] or an adjunct quantum field theory as proposed by

Wheeler and Feynman[4]. We would like to make a few remarks on these possibilities.

The adjunct field approach was developed in an attempt to eliminate the theoretically unsatisfactory necessity for employing Renormalization Theory in order to remove the infinite self-energies in Quantum Electrodynamics. One difficulty encountered by Feynman in quantizing their theory was that he discovered that the 'energy' levels were complex[7]. We believe that this is explained because he employed the potentials which do not account for the energy in transit and therefore in effect he employed $H_{\text{dynamical}} = \sum_a H_a(t)$ which does not correspond to the total conserved energy of the system. It is suggested that perhaps the potentials augmented by the expressions for the energy in transit should be employed to construct the adjunct fields.

The above suggestion that $H_{\text{conserved}} = p^\circ c$ is the quantity to focus upon is not only motivated by general principles, but also, more concretely, by the results of Andersen and von Baeyer[22] on the Bohr quantization of the two-body circular orbit solutions of Schild[23]. As they point out, their result coincides exactly with the 0(4, 2) infinite component wave equation approach of Barut and Baiquini[24] and Fronsdal[25] and with the Dirac formula for orbits with minimum eccentricity[26].

Any attempt to imitate Dirac[27] in order to find a relativistic wave equation in which $p_{ai} \to -g_{00}(\hbar/i)/(\partial/\partial x_a^i)$ immediately poses at least two major questions of interpretation. One is that $H_{\text{dynamical}} = p^\circ c$, where p° is given by Equation (12), when written in terms of the canonical variables $\{\mathbf{x}_a, \mathbf{p}_a\}_{a=1}^n$ raises, the question of the interpretation of the operators

$$\sqrt{1-(\mathbf{x}_a'/c)^2} = \frac{1}{\sqrt{1+\dfrac{1}{m_a^2 c^2}\left(\dfrac{\hbar}{i}\nabla_a - \dfrac{e_a}{l}\mathbf{A}_a\right)^2}}, \tag{35}$$

where $\mathbf{A}_a$ now also depends on the particle gradients. The other major question is how the advanced and retarded times are to be dealt with in such an approach.

Finally, we would like to point out that the observability of the canonical momenta follows, since the arguments of Aharonov and Bohm[28] concerning the observability of the electromagnetic potentials $\mathbf{A}_a$ are applicable in any quantum version of this theory.

ACKNOWLEDGEMENTS

In addition to the acknowledgements to Professors M. Flato and O. Navaro in the text, one of us (R.L.A.) would like to thank Professors R. Rączka and I. Białnicki-Burula for initiating his interest in the theory of directly interacting particles. The authors wish to express their deepest gratidude to Professor Plebański for his constant encouragement, many discussions of his results and insights about the theory of directly interacting particles, and in particular for his emphasis on the role of the Poisson bracket as a fiducial mark in developing this theory. In the latter regard one of us (R.L.A.) would like to acknowledge a discussion with Z. Iwinski. One of us (J.G.N.) wishes to acknowledge Professor A. Schild for several illuminating discussions on this work.

The authors are very grateful to Professor N. Svartholm for the hospitality and support extended to them at the Institute of Theoretical Physics, CTH, in Gothenburg, Sweden where this work began and to the Research Corporation for a grant which made the conclusion of the work possible. One of us (J.G.N.) wishes to express his sincere thankfulness to Professor J. Bach Andersen and his colleagues at Elektroafdelingen, Ingeniorakademiet for the kind hospitality he received at the department.

University of the Pacific, Stockton, Calif.

Danmarks Ingeniørakademi, Aalborg, Denmark

NOTES AND REFERENCES

[1] Schwarzschild, K., *Göttinger Nachrichten* **128** (1903), 132.

[2] Tetrode, H., *Z. Phys.* **10** (1922), 317.

[3] Fokker, A. D., *Z. Phys.* **58** (1929), 386; *Physica* **9** (1929), 33; **12** (1932), 145.

[4] Wheeler, J. A. and Feynman, R. P., *Rev. Mod. Phys.* **17** (1945), 157; *Rev. Mod. Phys.* **21** (1949), 425.

[5] Dirac, P. A. M., *Proc. Roy. Soc. (London)* **A167** (1938), 148.

[6] Gupta, Suraj N., *Proc. Phys. Soc. (London)* **A64** (1951), 50.

[7] Feynman, R. P., 'Nobel Lecture', *Physics Today* **19** (1966), 31.

[8] We are indebted to Professor M. Flato for emphasizing the importance of establishing the observability of the canonical momenta.

[9] The indices a and b denote particles. The quantities $x_a = \{x_a^\mu \mid \mu = 1, 2, 3, 0\}$, m_a, and e_a correspond to particle a's space-time coordinates measured in cm, rest mass measured in

gms and charge measured in franklins, respectively. The metric tensor is given by:

$$g_{\mu\nu}=g_{00}\begin{pmatrix} -1 & 0 & 0 & 0 \\ 0 & -1 & 0 & 0 \\ 0 & 0 & -1 & 0 \\ 0 & 0 & 0 & 1 \end{pmatrix}, \quad \text{and} \quad x=x^{\mu}g_{\mu\nu}x^{\nu}=x^{\mu}x_{\nu}.$$

The quantity $x'^{\mu}(t)$ is equal to the usual-time derivative of $x_a^{\mu}(t)$. The primed sum $\sum'^{n_1}_{a,b=1}$ corresponds to the sum over all pairs of distinctly different particles.

[10] Dettman, J. W. and Schild, A., *Phys. Rev.* **95** (1954), 1057.

[11] See e.g. Rzewuski, Jan, *Field Theory*, Part I, PWN-Polish Scientific Publishers, Warsaw, 1964.

[12] Bergmann, P.G., *Introduction to the Theory of Relativity*, Prentice-Hall, Inc., New York, 1942.

[13] The 4-momentum conservation law for this theory was first derived employing physical arguments by Wheeler and Feynman[4]. Dettman and Schild[10] later derived the full set employing symmetry arguments and a mathematical identity for double integrals. We would like to point out that in the latter case, it is necessary to supplement their argument with an argument that their recasting of the double integrals is unique. This e.g. is not so in the case of single integrals of functions of one variable.

[14] See e.g. the article by Peter Havas on 'Galilei- and Lorentz-Invariant Particle Systems and Their Conservation Laws', in *Problems in the Foundations of Physics* (ed. by Mario Bunge), vol. 4, Springer, Berlin-Heidelberg-New York, 1962.

[15] The present authors have formally generalized the main result of this paper, i.e. the inversion of the equations $p_{ai}=\partial L_a/\partial x_a'^i$ via the mathematical device of an integral operator to a class of particle systems characterized by the requirements that: (i) $\sum_{a=1}^{n}\int_{t_1}^{t_2}\mathrm{d}t\,\mathscr{L}_a(t)$ and $\sum_{a,b=1}^{n}\int_{t_1}^{t_2}\mathrm{d}t\int_{t_1}^{t_2}\mathrm{d}t'\,\mathscr{L}_{a,b}(t,t')$ are each Poincaré invariant and (ii) $\mathscr{L}_a(t)$ and $\mathscr{L}_{ab}(t,t')$ are each positive homogeneous functions of degree one separately in each set of particle 4-velocities $\{x_a'^{\mu}\}$. This includes e.g. the systems proposed by Havas, P. and Plebański, J., *Am. Phys. Soc.* **5** (1960), 433 and Van Dam, H. and Wigner, E. P., *Phys. Rev.* **142** (1966), 838.

[16] We sketch here a general method of solving Equation (23) to any prescribed order in the product of the various η_{ab}'s. Assume that we wish to solve Equation (31) up to and including the Nth order. First, retain only those terms up to and including I^{N-1}, then iterate on this truncated power series and discard any term which is of any order in the η_{ab}'s greater than N. This mode of solution leads to truncated 'Jacobs ladder'[7] diagrams.

[17] We are indebted to Professor M. Flato for this observation and to Professor O. Navaro for pointing out that our results are consistent with the results of R. P. Feynman[7] for the nonconservation of his associated quantum mechanical Hamiltonian operator for this problem and those of I. Prigogine and his colleagues (see e.g., *Chemica Scripta* **4** (1973), 5.

[18] Einstein, A., *Ann. Physik* **17** (1905), 891.

[19] Einstein, A., *Ann. Physik* **17** (1905), 132.

[20] de Broglie, L., *Comptes Rendus* **177** (1923), 507, 548, 630; *Nature* **112** (1923), 540, These de doctorat, Masson, Paris, 1924.

[21] Schrödinger, E., *Ann. Physik* **79** (1926), 361, 489; **80** (1926), 437; **81** (1926), 109.

[22] Andersen, C. M. and von Baeyer, Hans C., *Annals of Physics* **60** (1970), 67.

[23] Schild, A., *Phys. Rev.* **131** (1963), 2762.

[24] The expression for the relativistic energy spectrum in the approach of Andersen and von Baeyer was first derived by Barut, A. O., and Baiquini, A., (*Phys. Rev.* **184** (1969), 1342).

[25] Fronsdal had previously shown that this infinite-component wave equation in the nonrelativistic limit is capable of yielding in a special case the exact nonrelativistic H-atom spectrum (*Phys. Rev.* **171** (1968), 1811).

[26] See e.g. Bethe, H. A. and Salpeter, E. E., *Handbuch der Physik* (ed. by S. Flugge), Vol. XXXV, p. 170, Springer, Berlin 1957.
[27] Dirac, P. A. M., *Proc. Roy. Soc. (London)* **A117** (1928), 610; **A118** (1928), 351; *The Principles of Quantum Mechanics*, 4th ed., Oxford University Press, London, 1958.
[28] Aharonov, Y. and Bohm, D., *Phys. Rev.* **115** (1959), 485.

TAKEHIKO TAKABAYASI

THEORY OF ONE-DIMENSIONAL RELATIVISTIC ELASTIC CONTINUUM FOR THE MODEL OF PARTICLES AND RESONANCES*

1. Introduction

Motivated by the desire to restore a concrete picture for elementary particles in the frame of space-time in accordance with the tradition of Cartesian clarity, de Broglie and his group explored the possibility of classifying and of synthetizing various elementary particles by a structure extended in Minkowski space. Actually they focussed upon the rotational motion abstracted from the internal movement of the supposed extended structure and constructed 'relativistic rotator model' [1, 2] an appealing property of which is the feasibility of explaining spin and isospin including half integer eigenvalues.

More recently a relativistic extended model for baryons and mesons has also been explored along the line in which Yukawa's bilocal model is generalized to a multilocal model, which performs dilatations and vibrations as well. This model has proved to have favorable features, such as linear trajectories and reasonable form factors in agreement with experiment [3, 4].

Generally speaking a relativistic extended system cannot be a rigid body, because of relativity, but must be a certain elastic system with internal cohesion in order to remain a finite continuum in course of its movement. Thus its size must depend on the state of motion and may vary during the motion. A simple and primitive example of such a relativistic extended model should be a one-dimensional elastic continuum ('string'). Indeed this is in a sense the most natural extension of the traditional point model which has hitherto been the basis of the usual theory of elementary particles. Recently it has been shown [5, 6] that such a 'string' taken as a model of hadrons has the capability of deriving the Veneziano amplitude and its multiparticle generalization, which had just been proposed from S-matrix point of view to satisfy the concept of 'duality' for hadronic reactions [7]. The duality between direct channel resonance formation and Regge pole exchange is an important concept and might be compared

M. Flato et al. (eds.), Quantum Mechanics, Determinism, Causality, and Particles, 179–216. *All Rights Reserved. Copyright © 1976 by D. Reidel Publishing Company, Dordrecht-Holland.*

to the celebrated wave-particle duality in wave mechanics, but it thus becomes plausible to interpret the new duality just as a characteristic feature of a certain extended structure. Moreover the picture that rotations and vibrations of a relativistic string underlie the sequence of hadronic states is reminiscent of de Broglie's imagination fifty years ago [8] that the stationary material waves inside the atom underlie the observed discrete energy levels of atoms.[1]

Now the theory of the relativistic string was originally formulated directly at quantum-mechanical level by the set of a wave equation and an infinite number of subsidiary conditions, or equivalently by a super-wave equation (called detailed wave equation) [9]. It is, however, relevant to reconstruct the theory by first establishing the classical theory of a relativistic string explicitly and then quantisizing it, because this makes the correspondence-theoretic foundation and realistic interpretation of the theory clearer, even though the final ingredients of the theory are the same as in the original one. In fact some of the subtle points in the theory of relativistic string lie in the classical level of theory, and even the problem of 'ghost' has its classical root. Though such reconstructions of the relativistic string theory have been investigated by several authors [10–15], we construct the theory in the present paper in a more general and unifying framework and give deeper analysis of the theoretical foundation. Since we need to treat a relativistic movement performing a finite deformation our string theory cannot be immediately surmised from the conventional theory of non-relativistic string. Strangely, it does not seem that such a theory had been elaborated by relativists even at the classical level.

We distinguish between '*realistic* viewpoint' and '*geometric* viewpoint'. First we establish the string theory in the realistic viewpoint, where each elementary constituent (say 'parton') of the string has three degrees of freedom. The most general and flexible representation of this theory is given in the 'partial general-covariant form'. Then by choosing a suitable gauge we obtain the 'Lorentz-covariant formalism'. Taking another gauge we get the equivalent non-covariant formalism which directly exhibits the physical meaning of theory. Next we consider the string theory in the geometric viewpoint. The model obtained differs from the realistic model above mentioned because here each 'parton' has only two degrees of freedom and represents the case of vanishing intrinsic parton mass.

The relationship between the 'realistic' and 'geometric' models is quite analogous to the relationship between the de Broglie-Proca field for massive 'photon' (or vector meson) [16] and the Maxwell field. We show that the geometric model can be regarded as the singular limiting case of the realistic model. The internal motion of the latter can be restricted to transversal ones by a certain physical constraint, and then we have a smooth passage from the realistic to the geometric models. We classify and explicitly illustrate actual movements of the relativistic string, to verify concrete picture. We find that our string model implies a striking realization of 'relativistic rotator model' by its states on the leading trajectory as well as by some other states.

Finally we briefly state about quantization and interactions, and also give some general remarks.

2. Realistic viewpoint

A motion of a finite string traces a two-dimensional timelike world strip bounded in the spacelike direction but unbounded in the timelike direction. We represent it as

$$x_\mu = x_\mu(\sigma, \tau), \quad \mu = 1, 2, 3, 0 \tag{2.1}$$

with the aid of two independent parameters σ and τ, where σ runs over a finite domain $[\sigma_0, \sigma_1]$ while τ over $(-\infty, \infty)$. The geometrical quantities intrinsic to the world strip are the line element on it $(\mathrm{d}x_\mu)^2 = G^{\alpha\beta}\, \mathrm{d}\zeta_\alpha\, \mathrm{d}\zeta_\beta$ and the surface element $\sqrt{-D_0}\, \mathrm{d}\sigma\, \mathrm{d}\tau$, where

$$D_0 = G^{11}G^{00} - (G^{10})^2; \qquad G^{\alpha\beta} = g_{\mu\nu}(\partial x^\mu/\partial\zeta_\alpha)(\partial x^\nu/\partial\zeta_\beta),$$
$$(\zeta_1 \equiv \sigma, \zeta_0 \equiv \tau, \quad \alpha, \beta = 1, 0).$$

For the Minkowski metric $g_{\mu\nu}$ we use $g_{\mu\nu} = \delta_{\mu\nu}(1, 1, 1, -1)^2$. $G^{00} = (\partial x_\mu/\partial\tau)^2$, $G^{11} = (\partial x_\mu/\partial\sigma)^2$ and $G^{10} = (\partial x_\mu/\partial\sigma)(\partial x^\mu/\partial\tau)$ are the first fundamental quantities of the surface, forming the fundamental tensor of a 2-dimensional Riemannian space, and have to satisfy

$$D_0 \leqq 0, \tag{2.2}$$

because the strip is timelike. At any point on the strip there exist two light-like directions, lying on the strip, given by

$$d\tau/d\sigma = (-G^{10} \pm \sqrt{-D_0})/G^{00}. \tag{2.3}$$

It is at first important to distinguish between the 'geometric viewpoint' and the 'realistic viewpoint'. In the former the world strip alone has physical sense such that one world strip is regarded to represent one and the same motion of the string, whereas in the latter viewpoint one and the same world strip corresponds to many different motions of the string corresponding to the situation that the same strip is woven by world lines differently. Henceforth we take this realistic viewpoint. (For geometric viewpoint see Section 5.) Then σ in (2.1) means the parameter to label each elementary constituent of the string, and hence it must always be a Lorentz scalar. The strip is interpreted as a bunch of world lines each corresponding to $\sigma = \text{const.}$ (This interpretation assumes that constituents are distinguishable at least in classical theory.) To preserve this meaning of σ the arbitrariness of the parameters (σ, τ) is limited within the transformation [15]

$$\sigma \to \sigma'(\sigma), \qquad \tau \to \tau'(\sigma, \tau). \tag{2.4}$$

Even more specifically we may consider the string as the $N \to \infty$ limit of the linear multilocal model which consists of N spacetime points $x_\mu^{(\alpha)}$ $(\alpha = 1, \ldots, N)$ subject to 'relativistic Hooke potential' between neighbours [6, 4]. Then σ is the relabelling of α by $\sigma = (\sigma_1 - \sigma_0)\alpha/N + \sigma_0$, and we can interpret that $d\sigma/(\sigma_1 - \sigma_0)$ is the *relative particle number* (or say 'parton number') contained in $d\sigma$. In so far as we keep this physical meaning of σ, σ is essentially fixed and we are not allowed to make an arbitrary transformation on σ. Therefore we agree to take σ as dimensionless and as $0 \leqq \sigma \leqq \pi$. On the other hand τ is an arbitrary parameter to specify a point along each world line. Thus any observable quantities and physical relations must be invariant under the 'partial general transformation'

$$\tau \to \tau'(\sigma, \tau), \qquad (\sigma' = \sigma). \tag{2.5}$$

Note that τ need not be a Lorentz scalar and that (2.5) contains a transformation which alters the dimension of τ.

It is important to note that the arbitrary parameter τ is also a redundant parameter so that it must be possible to eliminate τ without leading to any ambiguity in the physical interpretation [15]. To see this we rewrite (2.1) as $x_i = x_i(\sigma, \tau)$ and $x_0 = x_0(\sigma, \tau) = t$. Then by solving the latter and inserting the result into $x_i = x_i(\sigma, \tau)$ we get the ordinary (non-covariant) representation for the motion of a string

$$x_i = x_i(\sigma, \tau(\sigma, t)) = x_i(\sigma, t), \tag{2.6}$$

which is unique in so far as the arbitrariness of (σ, τ) is restricted within (2.4) [15]. The ordinary velocity V_i of a certain element of the string is given as

$$V_i(\sigma, t) = \frac{\partial x_i(\sigma, t)}{\partial t} = \left(\frac{\partial x_i(\sigma, \tau)}{\partial \tau}\right) \Big/ \left(\frac{\partial x_0(\sigma, \tau)}{\partial \tau}\right). \tag{2.7}$$

Then causality requires that

$$\left|\frac{\partial \mathbf{x}(\sigma, t)}{\partial t}\right| \leqq 1, \quad \text{i.e.} \quad G^{00} = \left(\frac{\partial x_\mu}{\partial \tau}\right)^2 \leqq 0 \quad \text{everywhere}. \tag{2.8}$$

(We employ the unit system where $c = 1$.) The condition (2.8) is invariant under (2.4). It is proved that if (2.8) holds, (2.2) is always valid, whence

$$G^{11} \geqq (G^{10})^2/G^{00}, \qquad ((G^{10})^2/G^{00} \leqq 0). \tag{2.9}$$

The 4-velocity is clearly $U_\mu = (-G^{00})^{-1/2}(\partial x_\mu/\partial \tau)$. Now, if the condition (2.8) is slightly more stringent such that $G^{00} < 0$, then we get $(\partial x_0/\partial \tau)^2 > 0$ everywhere, and by reason of continuity $\partial x_0/\partial \tau$ must have a definite sign: that is, $\partial x_\mu/\partial \tau$ is either forward timelike everywhere, or backward timelike everywhere. Next the ordinary 3-diemensional radius vector going from the position of an element σ to that of $\sigma + d\sigma$ of the string, viewed at time t by a Lorentz observer is

$$\frac{\partial x_i(\sigma, t)}{\partial \sigma} d\sigma \equiv W_i(\sigma, t)\, d\sigma = \left(\frac{\partial x_i(\sigma, \tau)}{\partial \sigma} - V_i \frac{\partial x_0(\sigma, \tau)}{\partial \sigma}\right) d\sigma. \tag{2.10}$$

$$(\partial V_i(\sigma, t)/\partial \sigma = \partial W_i(\sigma, t)/\partial t).$$

The length along the string from the end $\sigma = 0$ to the point σ, viewed by Lorentz observer is

$$l(\sigma, t) = \int_0^\sigma \sqrt{\mathbf{W}^2}\, d\sigma. \tag{2.11}$$

In the present realistic viewpoint V_i and W_i are observables and are indeed left invariant under (2.5).

Any physically realizable motion has to satisfy a certain equation of motion and boundary condition on $x_\mu(\sigma, \tau)$, but they are physically meaningful only if they are covariant under (2.5), and therefore they should be derived from an action integral which is invariant under (2.5) (as well as under the Poincaré group). We find that for free cases such an action is restricted to the form

$$A=\int_{\tau_0}^{\tau_1} \mathrm{d}\tau \int_0^\pi L\, \mathrm{d}\sigma, \qquad L=-\kappa\sqrt{-D_\omega}, \tag{2.12a}$$

where

$$D_\omega=G^{00}(G^{11}+\omega)-(G^{01})^2=D_0+\omega G^{00}, \qquad (\omega\geqq 0). \tag{2.12b}$$

[We discard the extremely complicated possibility.] Note that due to the causality condition (2.8) we have $D_\omega=D_0+\omega G^{00}\leqq 0$ in so far as $\omega\geqq 0$. The above L contains two structure constants κ and ω with the dimensions $[ML^{-1}]$ and $[L^2]$, respectively. In place of them we may employ the constants

$$\mu_0=\kappa\sqrt{\omega}, \qquad K=\kappa/\sqrt{\omega}, \tag{2.13}$$

with the dimension $[M]$ and $[ML^{-2}]$, respectively. They are related to the modulus of internal tension and the mass density of the string (cf. Section 4), and L is rewritten as

$$L=-\mu_0\sqrt{-\left(G^{00}+\frac{K}{\mu_0}D_0\right)}. \tag{2.14}$$

(2.5) and (2.14) indicate that the theory is natural generalization of the relativistic point mechanics, where the free equation of motion must be derived form the action integral $W=-m_0\int\sqrt{-(\mathrm{d}x_\mu/\mathrm{d}\tau)^2}\,\mathrm{d}\tau$ which is invariant under arbitrary transformation $\tau\to\tau'(\tau)$. As seen later the two structure constants determine the trajectory slope and the leading intercept by

$$\alpha'=(2\pi\hbar\kappa)^{-1}, \qquad \alpha(0)=-\pi\kappa\omega/(2\hbar).$$

[See Equation (3.23).] We assume that the constant κ is a *universal constant* with the value

$$\kappa \approx 0.4 \text{ g s}^{-1}. \tag{2.15}$$

Though this κ does not occur in the fundamental equations in *free* case, it defines the scale of momentum and angular momentum.

Now the variational principle that $\delta A = 0$ with respect to an arbitrary variation $\delta x_\mu(\sigma, \tau)$ vanishing at initial τ_0 and at final τ_1 yields the Euler equation and the boundary condition

$$\frac{\partial p_\mu}{\partial \tau} = \frac{\partial S_\mu}{\partial \sigma}, \qquad S_\mu \big|_{0,\pi} = 0, \tag{2.16a, b}$$

where

$$\begin{aligned} p_\mu &= \frac{\partial L}{\partial x^\mu{}_{,0}} = \kappa \frac{(G^{11} + \omega)\, x_{\mu,0} - G^{10} x_{\mu,1}}{\sqrt{-D_\omega}}, \\ S_\mu &= -\frac{\partial L}{\partial x^\mu{}_{,1}} = \kappa \frac{G^{10} x_{\mu,0} - G^{00} x_{\mu,1}}{\sqrt{-D_\omega}}, \end{aligned} \tag{2.17}$$

and we used the notations

$$x_{\mu,0} \equiv \frac{\partial x_\mu(\sigma, \tau)}{\partial \tau}, \qquad x_{\mu,1} \equiv \frac{\partial x_\mu(\sigma, \tau)}{\partial \sigma}.$$

The following identities hold:

$$S^\mu x_{\mu,0} = 0, \qquad x_{\mu,0} p^\mu - L = 0. \tag{2.18a, b}$$

Thus (2.16b) implies three conditions for each end.

Unless $\omega = 0$,[3] we can reduce (2.16a) to the following form of equation of motion

$$\xi_{\mu\nu} \left[(G^{11} + \omega) \frac{\partial^2 x^\nu}{\partial \tau^2} - 2G^{10} \frac{\partial^2 x^\nu}{\partial \tau \partial \sigma} + G^{00} \frac{\partial^2 x^\nu}{\partial \sigma^2} \right] = 0, \tag{2.19}$$

where $\xi_{\mu\nu}$ is

$$\xi_{\mu\nu} = g_{\mu\nu} - \frac{\partial x_\mu}{\partial \tau} \frac{\partial x_\nu}{\partial \tau} \Big/ G^{00}, \qquad \left(\xi_{\mu\nu} \frac{\partial x^\nu}{\partial \tau} = 0 \right).$$

Clearly (2.19) is invariant under the 'partial general transformation' (2.5) and contains *three* independent equations. This corresponds to the fact that each element of the string has three degrees of freedom.

Equation (2.16) implies the 4-momentum conservation law which holds on the world strip *locally.* Thus the 4-momentum of the system is given by the line integral

$$P_\mu = \int_C (p_\mu \, d\sigma + S_\mu \, d\tau), \tag{2.20}$$

where C is an arbitrary curve on the strip going from the end $x_\mu(0, \tau)$ at any τ to the other end $x_\mu(\pi, \tau')$ at any τ'. (2.20) does not depend on the choice of C, because $p \, d\sigma + S \, d\tau$ is a total differential due to (2.16a) and because moreover we have (2.16b). Taking any equal τ curve on the strip for C we get

$$P_\mu = \int_0^\pi p_\mu(\sigma, \tau) \, d\sigma, \qquad dP_\mu/d\tau = 0. \tag{2.21}$$

Similarly, the angular momentum tensor is

$$M_{\mu\nu} = \int_C (x_{[\mu} p_{\nu]} \, d\sigma + x_{[\mu} S_{\nu]} \, d\tau), \tag{2.22}$$

which again does not depend on C and is written as

$$M_{\mu\nu} = \int_0^\pi x_{[\mu}(\sigma, \tau) \, p_{\nu]}(\sigma, \tau) \, d\sigma, \qquad dM_{\mu\nu}/d\tau = 0. \tag{2.23}$$

P_μ and $M_{\mu\nu}$ are invariant under (2.5) and are observables.

$p_\mu(\sigma, \tau)$ is also the canonical conjugate to $x_\mu(\sigma, \tau)$, but it is verified that there exists one local identity between p_μ and $\partial x_\mu/\partial\sigma$:

$$\left(p \frac{\partial x}{\partial \sigma}\right)^2 + \omega \left\{ p^2 + \kappa^2 \left[\left(\frac{\partial x}{\partial \sigma}\right)^2 + \omega \right] \right\} = 0, \tag{2.24}$$

which reflects the covariance of theory under (2.5).

3. Lorentz-covariant formalism

To proceed now to obtaining the solutions and later to quantization it is convenient to choose a suitable gauge in exploiting the invariance of (2.16) or (2.19) under (2.5). We have two particularly important gauges, the 'Lorentz-covariant' one and the 'Lorentz-noncovariant' one. In this section we consider the former, where we impose

$$G^{10}=0 \quad \text{i.e.} \quad U^{\mu}x_{\mu,1}=0. \tag{3.1}$$

This means to choose τ such that an equal τ curve on the strip is everywhere orthogonal to world lines. Such τ must now be a Lorentz scalar. (3.1) is reexpressed as

$$\partial x_0(\sigma,\tau)/\partial\sigma=(\mathbf{VW})/(1-\mathbf{V}^2). \tag{3.2}$$

Under (3.1), the inequality (2.9) which results from the causality condition simplifies to

$$G^{11}\geqq 0 \quad \text{everywhere.} \tag{3.3}$$

Originally the transformation (2.5) implied that τ's origin can be taken arbitrarily for each world line, but (3.3) now ensures that τ is an 'instant parameter' such that an equal τ curve on the strip is a space-like curve. Owing to this fact the definitions of the global quantities of the system, such as the geometrical center-of-mass X_{μ}, as integrals with respect to σ at equal τ (See (3.8)) become physically suitable.

From (2.19) and (3.1) we get $G^{11}+\omega=-\varphi(\tau)\,G^{00}$, where $\varphi(\tau)$ is an arbitrary function of τ, but satisfies $\varphi(\tau)>0$ (in so far as $\omega\geqq 0$) because of (3.3). Further Equations (2.17), (2.19) and (2.16b) simplify to

$$p_{\mu}=\kappa\sqrt{\varphi(\tau)}\,x_{\mu,0}, \qquad S_{\mu}=(\kappa/\sqrt{\varphi(\tau)})\,x_{\mu,1},$$
$$\frac{\partial^2 x_{\mu}}{\partial\sigma^2}=\varphi(\tau)\frac{\partial^2 x_{\mu}}{\partial\tau^2}+\frac{1}{2}\frac{d\varphi}{d\tau}\frac{\partial x_{\mu}}{\partial\tau}, \quad \text{and} \quad x_{\mu,1}\,|_{0,\pi}=0.$$

Next we use $\tilde{\tau}=\int^{\tau}\varphi(\tau)^{-1/2}\,d\tau$ in place of τ. Then the above equations simplify to $G^{11}+\omega=-(\partial x/\partial\tilde{\tau})^2$, $p_{\mu}=\kappa\cdot\partial x_{\mu}/\partial\tilde{\tau}$, and $\partial^2 x_{\mu}/\partial\tilde{\tau}^2=\partial^2 x_{\mu}/\partial\sigma^2$. Owing to this first relation and (3.1), the parameter $\tilde{\tau}$ is essentially uniquely fixed (aside of $\tilde{\tau}\to\tilde{\tau}+\delta$) so that it must have a definite physical meaning. Indeed we have now $(\partial x_{\mu}/\partial\tilde{\tau})^2|_{0,\pi}=-\omega$, so that $\sqrt{\omega}\,\tilde{\tau}$ means the proper time of the ends (unless $\omega=0$). Henceforth we write $\tilde{\tau}$ simply as τ. Thus finally the fundamental equations in this gauge consist of the d'Alembert equation and the open-end boundary condition

$$\frac{\partial^2 x_\mu}{\partial \tau^2} = \frac{\partial^2 x_\mu}{\partial \sigma^2}, \qquad \frac{\partial x_\mu}{\partial \sigma}\Big|_{0,\pi} = 0, \tag{3.4a, b}$$

and the constraint

$$G^{00} + G^{11} = -\omega. \tag{3.5}$$

The remarkable feature of the present gauge is that it produces curious pseudosymmetry between σ and τ notwithstanding the asymmetry of the original Lagrangian (2.12) due to non-zero ω. This pseudo-symmetry means that the fundamental equations, apart from the boundary condition and the causality inequality, are invariant under the interchange $\sigma \to \tau$, $\tau \to \sigma$. Since σ and τ have different domains this 'symmetry' has local sense only. Another feature is that the equation of motion is originally nonlinear but, with the present constraint which itself is quadratic, becomes linear. Of course the superposition principle does not apply to physical solutions. In those connections we now mention some characteristic points of this formalism. [The results in the following apply to the $\omega - 0$ case also, where, however, caution must be payed as to interpretation (See Section 5).]

For the while we restrict our attention to linear equations (3.4a, b) alone (i.e. the equation of motion and boundary condition) in this gauge.

(i) They are derivable from the simpler action integral

$$A_1 = \iint \mathrm{d}\tau\, \mathrm{d}\sigma L_1, \qquad L_1 = \frac{\kappa}{2}(G^{00} - G^{11} - \omega), \tag{3.6}$$

which is numerically equal, under the constraint, to the original L.[4] Then we have the corresponding Hamiltonian density $H_1 = (\partial x/\partial \tau)\, p - L_1 = \kappa/2\, (G^{00} + G^{11} + \omega)$, which gives the scalar Hamiltonian

$$\bar{H}_1 = \Lambda^0 + \frac{\pi\kappa}{2}\,\omega, \qquad \Lambda^0 = \tfrac{1}{2}\int_0^\pi \left[\frac{1}{\kappa}\, p^2 + \kappa\left(\frac{\partial x}{\partial \sigma}\right)^2\right] \mathrm{d}\sigma. \tag{3.7}$$

This allows the canonical formalism in which the role of t in the usual case is taken over by τ, since the equation of motion (3.4) is reproduced via

$$\frac{\partial x_\mu}{\partial \tau} = \frac{\delta \bar{H}_1}{\delta p^\mu}, \qquad \frac{\partial p_\mu}{\partial \tau} = -\frac{\delta \bar{H}_1}{\delta x^\mu}.$$

(ii) The definition of center-of-mass depends on the gauge, but

$$X_\mu(\tau) = \frac{1}{\pi} \int_0^\pi x_\mu(\sigma, \tau)\, d\sigma \tag{3.8}$$

defined in the present gauge is an unambiguous quantity, which we shall call 'geometrical center-of-mass'. It satisfies

$$\frac{dX_\mu}{d\tau} = \frac{P_\mu}{\pi\kappa}, \qquad \frac{d^2 X_\mu}{d\tau^2} = 0,$$

so that $m\tau/(\pi\kappa)$ has the meaning of the proper-time of the center-of-mass. (This applies to the $\omega = 0$ case inclusive.)

(iii) The action integral A_1 and Equations (3.4a, b) have very great symmetry properties, which are not necessarily shared by the original action integral A. They are invariant under several 'external transformations' acting on x_μ as well as under several 'internal transformations' acting on (σ, τ). The translation and Lorentz groups, and also the external dilatation $x_\mu(\sigma, \tau) \to \lambda x_\mu(\sigma, \tau)$, and the (σ, τ)-dependent translation (meaning nonlinear deformation)

$$x_\mu(\sigma, \tau) \to x_\mu(\sigma, \tau) + d_\mu \cdot \begin{matrix} \cos r\tau \\ \cdot \\ \sin r\tau \end{matrix} \cos r\sigma, \tag{3.9}$$

$(d_\mu = \text{const vector},\ r = \text{integer})$

belong to the former. (Note, however, that under dilatation the action integral is not invariant.)

The internal transformation which leaves A_1, whence (3.4) invariant, consists of special linear transformations, which do not mix σ and τ, and the nonlinear conformal transformation, which mixes σ and τ and does not belong to (2.4). The former consists of the internal scale transformation ($\sigma' = \alpha\sigma$, $\tau' = \alpha\tau$, $\sigma_0' = 0$, $\sigma_1' = \alpha\pi$), τ-displacement ($\tau' = \tau + \delta$), internal reflection ($\sigma' = \pi - \sigma$), and τ-reversal. The transformation (2.4) is reduced to this smaller 'special linear group', and the causality condition (3.3) is preserved under this transformation. By conformal transformation we mean the one which leaves the relation $d\sigma^2 - d\tau^2 = 0$ invariant and satisfies $[\partial\tau'/\partial\sigma]_{0,\pi} = 0$.[5] This is given in terms of an arbitrary periodic function $f(\lambda)=\sum_n f_n e^{in\lambda}$, $(f_n^*=f_{-n})$, as

$$\begin{aligned}\sigma^c(\sigma,\tau)&=\sigma+\tfrac{1}{2}(f(\tau+\sigma)-f(\tau-\sigma))=\sigma+i\sum_n f_n e^{in\tau}\sin n\sigma,\\ \tau^c(\sigma,\tau)&=\tau+\tfrac{1}{2}(f(\tau+\sigma)+f(\tau-\sigma))=\tau+\sum_n f_n e^{in\tau}\cos n\sigma.\end{aligned}\tag{3.10}$$

The conformal symmetry implies that if we find a solution $x_\mu(\sigma,\tau)$ to the equation of motion and boundary condition (3.4) we obtain at once infinitely many other solutions $x_\mu^c(\sigma,\tau)$ given by $x_\mu(\sigma^c(\sigma,\tau),\tau^c(\sigma,\tau))= =x_\mu^c(\sigma,\tau)$.

We now mention the conservation laws related to the above symmetries. The conservation laws of 4-momentum and angular momentum tensor hold of course. They are (2.20) and (2.22), where now $p_\mu=\kappa(\partial x_\mu/\partial\tau)$ and $S_\mu=\kappa(\partial x_\mu/\partial\sigma)$. The invariance under (3.9) corresponds to the existence of an infinite number of conserved vectors

$$C_\mu^r=\frac{1}{\sqrt{2}\pi}e^{ir\tau}\int_0^\pi\left(\frac{\partial x_\mu}{\partial\tau}\cos r\sigma+i\frac{\partial x_\mu}{\partial\sigma}\sin r\sigma\right)\mathrm{d}\sigma,\qquad (r=\text{integer})$$

each of which means the weighted amplitude of each normal mode except for $C_\mu^0=P_\mu/(\sqrt{2}\,\pi\kappa)$.

The invariance of A_1 under the conformal group gives an infinite number of constants of motion

$$\Lambda^r=e^{ir\tau}\Lambda^r(\tau)=\Lambda^r(0),\tag{3.11}$$

where

$$\begin{aligned}\Lambda^r(\tau)&=\int_0^\pi(F\cos r\sigma+iG\sin r\sigma)\,\mathrm{d}\sigma.\qquad (r=\text{integer})\\ F&=\frac{\kappa}{2}(G^{00}+G^{11}),\qquad G=\kappa G^{10}.\end{aligned}\tag{3.12}$$

Since the conformal transformation concerns the internal parameters it commutes with the Poincaré group acting on x_μ. Therefore P_μ and $M_{\mu\nu}$ are conformal invariant, while X_μ is not.

(iv) Equations (3.4a, b) imply that $\partial x_\mu(\sigma,\tau)/\partial\tau\pm\partial x_\mu(\sigma,\tau)/\partial\sigma$ are the same functions of $\tau\pm\sigma$ only, which we denote as $v_\mu(\tau\pm\sigma)$, and moreover that these functions are periodic:

$$v_\mu(\lambda+2\pi)=v_\mu(\lambda). \tag{3.13}$$

Further in terms of C_μ^r defined before this is expressed as

$$v_\mu(\lambda)=\sqrt{2}\sum_{r=-\infty}^{\infty} C_\mu^r\, e^{-ir\lambda}. \tag{3.14}$$

We denote the co-ordinate of the end as $u_\mu(\tau)\equiv x_\mu(0,\tau)$. Then $v_\mu(\tau)= = \mathrm{d}u_\mu(\tau)/\mathrm{d}\tau$. $u_\mu(\tau)$ and $v_\mu(\tau)$ represent the motion of the end and are very useful quantities [17], because a motion of the whole string is completely specified by the motion of the end. (See below.) We have in particular

$$P_\mu=\frac{\kappa}{2}\int_{-\pi}^{\pi} v_\mu(\tau)\,\mathrm{d}\tau, \qquad \Lambda^r=\frac{\kappa}{4}\int_{-\pi}^{\pi} (v_\mu(\tau))^2\, e^{ir\tau}\,\mathrm{d}\tau. \tag{3.15}$$

(v) If we introduce the relative co-ordinate and momentum, $\bar{x}_\mu(\sigma,\tau)= =x_\mu(\sigma,\tau)-X_\mu(\tau)$ and $\bar{p}_\mu(\sigma,\tau)=p_\mu(\sigma,\tau)-P_\mu/\pi$, then $\bar{x}_\mu(\sigma,\tau)$ is periodic in τ, and $M_{\mu\nu}$ and Λ^0 also split into the part due to the center-of-mass motion and the part due to the internal motion such that $M_{\mu\nu}=X_{[\mu}P_{\nu]}+S_{\mu\nu}$, $\Lambda^0= =[1/2\kappa(P_\mu)^2+R]/\pi$, where $S_{\mu\nu}$ and R, too, are constants of motion. (See Ref. [15].) The magnitude of spin (i.e. the intrinsic angular momentum) is given (classically) by $J=\sqrt{-(\omega_\mu)^2/(P_\mu)^2}$ with $\omega_\mu=\frac{1}{2}\varepsilon_{\mu\nu\kappa\lambda}M^{\nu\kappa}P^\lambda= =\frac{1}{2}\varepsilon_{\mu\nu\kappa\lambda}S^{\nu\kappa}P^\lambda$.

We now take account of the constraints (3.1) and (3.5). It is clear that the constraint breaks the conformal invariance and also the external dilatation symmetry, unless $\omega=0$. The constraint is also represented as

$$(v_\mu(\tau))^2=-\omega. \tag{3.16}$$

This means that the end moves causally and this guarantees that every point of the string also moves causally with *three* degrees of freedom. As remarked before (3.16) means that $\sqrt{\omega}\,\tau$ is the proper time of the ends. Moreover we can prove conversely that the condition that τ should mean the proper time of the end is equivalent to the constraint (3.1), via the equation of motion and boundary condition. (3.16) also gives the relation $(P_\mu)^2\leqq-\pi^2\mu_0^2$, so that no tachyonic motion occurs and the mass m satisfies $m\geqq\pi\mu_0$.

The constraint (3.16) is also expressed as

$$\Lambda^0 = -\frac{\pi\kappa}{2}\,\omega, \qquad \Lambda^r = 0 \qquad (r = \pm 1, \pm 2, \ldots). \tag{3.17a, b}$$

(3.17a) means that the squared mass is

$$m^2 = -(P_\mu)^2 = 2\kappa R + \pi^2 \mu_0^2. \tag{3.18}$$

Also (3.17) implies that the conservation laws of Λ^r are apparent ones, and Λ^r, in contrast to C^r_μ, are not those constants of motion that classify physical solutions.

The fact that under the constraints the parameters σ and τ are essentially unique corresponds to the fact that they have definite physical meanings, as already explained. They still have slight arbitrariness within the special linear group, under which all fundamental equations are invariant. In any case, however, in the present gauge all $\partial x_\mu/\partial\tau$ and $\partial x_\mu/\partial\sigma$ can be expressed completely in terms of physical quantities $V_i(\sigma, t)$ and $W_i(\sigma, t)$. Indeed by using the constraints we have, besides (3.2),

$$\frac{\partial x_0(\sigma,\tau)}{\partial\tau} = \frac{\sqrt{-G^{00}}}{\sqrt{1-\mathbf{V}^2}}, \qquad \frac{\partial x_i(\sigma,\tau)}{\partial\sigma} = W_i + \frac{(\mathbf{VW})}{1-\mathbf{V}^2}\,V_i,$$

$$-G^{00} = \mathbf{W}^2 + \frac{(\mathbf{VW})^2}{1-\mathbf{V}^2} + \omega, \quad \text{etc.} \tag{3.19}$$ [6]

The present gauge is the most convenient for writing down the general solutions. In fact there are various ways to express the general solutions, each of which is useful.

(i) One can give it in the form of the Cauchy problem. For that purpose we first consider that solution of

$$\frac{\partial^2 D}{\partial\tau^2} - \frac{\partial^2 D}{\partial\sigma^2} = 0, \qquad \left.\frac{\partial D}{\partial\sigma}\right|_{0,\pi} = 0,$$

which satisfies the initial conditions

$$\left.\frac{\partial D}{\partial\tau}\right|_{\tau=0} = \delta(\sigma), \qquad D(\sigma, 0) = 0.$$

This is given by

$$D(\sigma,\tau)=\tfrac{1}{4}\sum_{n=-\infty}^{\infty}[\varepsilon(\tau-\sigma-2n\pi)+\varepsilon(\tau+\sigma-2n\pi)]=$$
$$=\tfrac{1}{2}\varepsilon(\tau)\sum_{n=-\infty}^{\infty}\theta(\tau^2-(\sigma+2n\pi)^2),$$

which is odd in τ and vanishes in the region where $|\tau|<\sigma$. Next we define $\Delta(\sigma,\sigma',\tau)=D(\sigma-\sigma',\tau)+D(\sigma+\sigma',\tau)$, which is even and symmetric in σ and σ' and has the properties that

$$\left.\frac{\partial\Delta(\sigma,\sigma',\tau)}{\partial\tau}\right|_{\tau=0}=\delta(\sigma-\sigma')+\delta(\sigma+\sigma')+\delta(2\pi-\sigma-\sigma'),$$
$$\Delta(\sigma,\sigma',\tau)=0\quad\text{for}\quad|\tau|<|\sigma-\sigma'|. \tag{3.20a, b}$$

Then the solution to (3.4) is expressed as

$$x_\mu(\sigma,\tau)=\int_0^\pi d\sigma'\Delta(\sigma,\sigma',\tau-\tau_0)\frac{\overleftrightarrow{\partial}}{\partial\tau_0}x_\mu(\sigma',\tau_0). \tag{3.21}$$

Due to the property of Δ, $x_\mu(\sigma',\tau_0)$ in the region $|\sigma-\sigma'|>|\tau-\tau_0|$ on the strip does not contribute to $x_\mu(\sigma,\tau)$. In the neighborhood of $x_\mu(\sigma,\tau)$ the causal relation exists only for the region $d\tau^2-d\sigma^2\geqq 0$, which means the timelike region because we have $[x_\mu(\sigma+d\sigma,\tau+d\tau)-x_\mu(\sigma,\tau)]^2=G^{00}$ $d\tau^2+G^{11}\,d\sigma^2=G^{00}(d\tau^2-d\sigma^2)-\omega\,d\sigma^2\leqq 0$. This verifies causality.

The condition (3.1) is incorporated in (3.21) by assuming that $G^{10}=0$ and $\partial G^{10}/\partial\tau=0$ at $\tau=\tau_0$.

(ii) A more simple way is suggested from the '$\sigma-\tau$ symmetry'. Namely we exchange the roles of τ and σ in (i) and regard the problem as a 'Cauchy problem' with respect to σ. Indeed the whole motion is determined once we are given the motion of the end $x_\mu(0,\tau)=u_\mu(\tau)$, which must have the property that its derivative $v_\mu(\tau)$ satisfies the conditions (3.13), and (3.16). In terms of such $u_\mu(\tau)$ the solution is given simply as $x_\mu(\sigma,\tau)=\frac{1}{2}[u_\mu(\tau-\sigma)+$ $+u_\mu(\tau+\sigma)]$, which in fact covers all possible solutions. Since the motion of the end $u_\mu(\tau)$ and $v_\mu(\tau)$ dominates the motion of the whole string, all relevant quantities are expressed in terms of it, as already remarked.

(iii) Another form of general solution is obtained by noting (3.14).

Namely

$$x_\mu(\sigma, \tau) = X_\mu(0) + \sqrt{2}\, C_\mu^0 \tau + \sqrt{2}i \sum_{r=-\infty}^{\infty}{}' \, C_\mu^r \, e^{-ir\tau} \cdot \frac{\cos r\sigma}{r}.$$

Corresponding to (3.17) C_μ^r must satisfy the constraints:

$$\tfrac{1}{2}(C^0)^2 + \sum_{n=1}^{\infty} C^{n*}C^n = -\frac{\omega}{4}; \qquad \sum_{n=-\infty}^{\infty} C^n C^{r-n} = 0, \qquad (r \neq 0).$$

Λ^r, R and $S_{\mu\nu}$ are rewritten as

$$\Lambda^r = \pi\kappa \sum_n C^n C^{r-n}, \qquad R = 2\pi^2\kappa \sum_{n=1}^{\infty} C^{n*}C^n,$$

$$S_{\mu\nu} = -2\pi i\kappa \sum_{n=1}^{\infty} C_{[\mu}^{n*} C_{\nu]}^n / n.$$

(iv) Another obvious way is to start from the expansion

$$x_\mu(\sigma, \tau) = \sum_{r=-\infty}^{\infty} x_\mu^r(\tau) \cos r\sigma, \qquad (x^{-r} = x^r)$$

$$p_\mu(\sigma, \tau) = \frac{1}{2\pi} \sum_{r=-\infty}^{\infty} p_\mu^r(\tau) \cos r\sigma, \qquad (p^{-r} = p^r)$$

where $x_\mu^r(\tau)$ satisfies $\mathrm{d}^2 x_\mu^r/\mathrm{d}\tau^2 + r^2 x_\mu^r = 0$, while $p_\mu^r(\tau) = 2\pi\kappa \cdot \mathrm{d}x_\mu^r(\tau)/\mathrm{d}\tau$. In particular $x_\mu^0(\tau) = X_\mu(\tau)$, $p_\mu^0 = 2P_\mu$. If we then define

$$C_\mu^r(\tau) = \frac{1}{\sqrt{2}} \left(\frac{\mathrm{d}x_\mu^r(\tau)}{\mathrm{d}\tau} - irx_\mu^r(\tau) \right), \quad C_\mu^{-r}(\tau) = C_\mu^r(\tau)^*,$$

the solution is $C_\mu^r(\tau) = C_\mu^r \, e^{-ir\tau}$, where C_μ^r are the same constants given before.

In the $m^2 - J$ plot the classical solutions are distributed continuously in the hatched region in Figure 1.

The solutions where only the first mode is excited besides C_μ^0 are simple and physically important. They are specified by the complex vector C_μ^1 which is restricted by

$$P^\mu C_\mu^1 = 0, \qquad (C_\mu^1)^2 = 0. \qquad \text{(3.22a, b)}$$

They give the solutions lying on the leading trajectory, which is given by

$$J=\frac{R}{\pi}=\frac{m^2}{2\pi\kappa}-\frac{\pi\kappa}{2}\,\omega. \tag{3.23}$$

By eliminating τ we see that the motion of the string for this solution is a rigid rotation and the string has constant length $\bar{l}=2\sqrt{2J/(\pi\kappa)}=(2/\pi\kappa)$

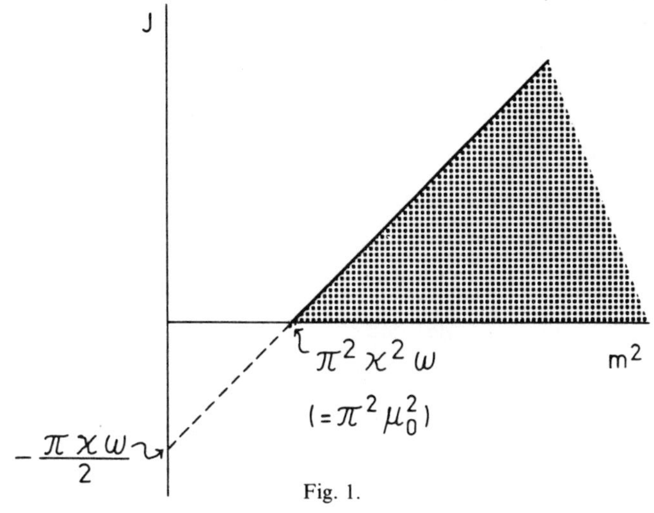

Fig. 1.

$[m^2-\pi^2\mu_0^2]^{1/2}$. The κ-value of (2.15) corresponds to $\alpha'\approx 1$ BeV^{-2}, and the above $\bar{l}$ means the extension of the order of nucleon Compton wavelength. If we assume that the dimensionless constant $\pi^2\kappa\omega/\hbar$ is of order 1, then we have $\omega\sim 1.6\times 10^{-27}$cm^2. In the ground state $(m=\pi\kappa\sqrt{\omega}=\pi\mu_0$, $J=0)$, however, the string shrinks up to a single point. (In quantum mechanics, however, the spatial extension of the ground state is not essentially different from that of the excited states due to quantum fluctuation.)

We may consider the limiting cases. In the tensionless limit $(K\to 0$, $\mu_0=\text{fixed}\neq 0)$ the leading trajectory becomes steeper to approach the vertical line (in Figure 1), with the ground state at $(m^2=\pi^2\mu_0^2, J=0)$ fixed. Then in a state on the leading trajectory $\bar{l}$ goes to ∞. The opposite is the local theory limit, $K\to\infty$ with $\mu_0=\text{fixed}\neq 0$, where the trajectory becomes flat and $\bar{l}\to 0$. Another limit is $\omega\to 0$, where the leading trajectory starts from the origin due to the external dilatation invariance. If $x_\mu(\sigma,\tau)$ is a solution at (m^2, J) then $\lambda x_\mu(\sigma,\tau)$ is a solution at $(\lambda^2 m^2, \lambda^2 J)$.

4. NON-COVARIANT FORMALISM

The theory of a realistic string model can also be represented in a Lorentz non-covariant formalism. This is obtained as follows. Since the original theory is covariant under arbitrary transformation (2.5), we can choose $\tau'(\sigma, \tau)$ such that it equals $x_0(\sigma, \tau)$ in the Lorentz frame under consideration, namely

$$\tau'(\sigma, \tau) = x_0(\sigma, \tau) = t. \tag{4.1}$$

This fixes the gauge completely. Now $x_i(\sigma, \tau)$ becomes $x_i(\sigma, \tau') = x_i(\sigma, t)$, and further we have

$$\begin{aligned} &\frac{\partial x_i(\sigma, \tau')}{\partial \tau'} = \frac{\partial x_i(\sigma, t)}{\partial t} = V_i, \qquad \frac{\partial x_0(\sigma, \tau')}{\partial \tau'} = 1, \\ &\frac{\partial x_i(\sigma, \tau')}{\partial \sigma} = \frac{\partial x_i(\sigma, t)}{\partial \sigma} = W_i, \qquad \frac{\partial x_0(\sigma, \tau')}{\partial \sigma} = 0. \end{aligned} \tag{4.2}$$

Thus in the present gauge $G^{\alpha\beta}$ and $-D_\omega$ become

$$\begin{aligned} &G^{00} \to -(1 - \mathbf{V}^2), \qquad G^{11} \to \mathbf{W}^2, \qquad G^{01} \to \mathbf{V}\mathbf{W}, \\ &-D_\omega \to \Delta_\omega = (\mathbf{W}^2 + \omega)(1 - \mathbf{V}^2) + (\mathbf{V}\mathbf{W})^2, \end{aligned} \tag{4.3}$$

so that the invariant action integral (2.12) is expressed as

$$A = -\kappa \int_{t_0}^{t_1} \mathrm{d}t \int_0^\pi \mathrm{d}\sigma \sqrt{\Delta_\omega}. \tag{4.4}$$

By (4.2) and (4.3) we see at once that the equation of motion (2.19) for the case $\omega \neq 0$ goes over in the present gauge to the form

$$(\mathbf{W}^2 + \omega)\frac{\partial^2 x_i}{\partial t^2} - 2\mathbf{V}\mathbf{W}\frac{\partial^2 x_i}{\partial t \partial \sigma} - (1 - \mathbf{V}^2)\frac{\partial^2 x_i}{\partial \sigma^2} = 0, \tag{4.5}$$

while the boundary condition (2.16b) becomes

$$W_i \big|_{\sigma = 0, \pi} = 0. \tag{4.6}$$

(4.5) and (4.6) also follow directly from the variational principle on (4.4) in so far as $\omega \neq 0$.[7] The fundamental Equations (4.5) and (4.6) in this gauge are Lorentz non-covariant but free from constraint.

If we denote p_μ in the present gauge as Π_μ, the 4-momentum of the system is

$$P_\mu = \int_0^\pi \Pi_\mu(\sigma, t)\, \mathrm{d}\sigma, \quad \text{with} \quad \frac{\mathrm{d}P_\mu}{\mathrm{d}t} = 0, \tag{4.7}$$[8]

so that Π_i and Π_0 mean the momentum and energy density per unit σ (i.e. per parton) *viewed by a Lorentz observer.*

From (2.17) and (4.2) they are

$$\Pi_i = \frac{\kappa}{\sqrt{\Delta_\omega}} [(\mathbf{W}^2 + \omega)\, V_i - (\mathbf{VW})\, W_i] = \kappa \frac{\partial(-\sqrt{\Delta_\omega})}{\partial V_i},$$
$$\Pi_0 = \kappa(\mathbf{W}^2 + \omega)/\sqrt{\Delta_\omega}. \tag{4.8}$$

Π_i is the momentum variable canonically conjugate to $x_i(\sigma, t)$, and $\mathbf{\Pi}$ is generally not parallel to $\mathbf{V}$. Also note that $\Pi_0 \geqq 0$ everywhere, by virtue of $|\mathbf{V}| \leqq 1$ and $\omega \geqq 0$, which mean causality. Likewise S_μ of (2.17) becomes

$$\tilde{S}_i = \frac{\kappa}{\sqrt{\Delta_\omega}} [(1 - \mathbf{V}^2)\, W_i + (\mathbf{VW})\, V_i] = \kappa \frac{\partial \sqrt{\Delta_\omega}}{\partial W_i},$$
$$\tilde{S}_0 = \kappa(\mathbf{VW})/\sqrt{\Delta_\omega}, \tag{4.9}$$

and the Euler equation is $\partial\Pi_\mu(\sigma, t)/\partial t = \partial\tilde{S}_\mu(\sigma, t)/\partial\sigma$. The relations (2.18) and (2.24) now take the forms

$$\tilde{S}_0 = \mathbf{V}\tilde{\mathbf{S}}, \qquad \Pi_0 = \mathbf{\Pi V} + \kappa\sqrt{\Delta_\omega}, \tag{4.10}$$

$$\Pi_0^2 = \mathbf{\Pi}^2 + \kappa^2\mathbf{W}^2 + \frac{1}{\omega}(\mathbf{\Pi W})^2 + \kappa^2\omega, \quad (\omega \neq 0). \tag{4.11}$$

Also we have the relations

$$\left.\frac{\partial \mathbf{\Pi}}{\partial \sigma}\right|_{0,\pi} = 0, \tag{4.12}$$

$$\mathbf{V} = [\mathbf{\Pi} + (\mathbf{\Pi W})\, \mathbf{W}/\omega]/\Pi_0, \qquad (\omega \neq 0). \tag{4.13}$$

In this gauge angular momentum tensor is written as[9]

$$M_{\mu\nu} = \int_0^\pi \mathfrak{M}_{\mu\nu}(\sigma, t)\, \mathrm{d}\sigma,$$
$$\mathfrak{M}_{ij} = x_{[i}\Pi_{j]}, \qquad \mathfrak{M}_{i0} = x_i\Pi_0 - t\Pi_i. \tag{4.14}$$

$\mathfrak{M}_{ij}$ means the angular-momentum density per parton in the Lorentz frame. We can verify the conservation law $\mathrm{d}M_{ij}/\mathrm{d}t = \mathrm{d}M_{i0}/\mathrm{d}t = 0$, which imply that the theory is Lorentz-invariant, even though the equation of motion is non-covariant. The 'physical center-of-mass' should be defined by

$$\tilde{X}_i(t) = \int_0^\pi x_i(\sigma, t)\, \Pi_0(\sigma, t)\, \mathrm{d}\sigma / P_0, \tag{4.15}$$

which conforms with the usual definition of center-of-mass of a relativistic extended system [18]. By (4.14), Equation (4.15) is re-expressed as

$$\tilde{X}_i = (P_i/P_0)\, t + M_{i0}/P_0, \tag{4.16}$$

which is also the usual relation.

In the present gauge the theory is put in the *standard* Hamiltonian formalism. Indeed from (4.11) the Hamiltonian density is

$$H = \Pi_0 = \left[\mathbf{\Pi}^2 + \kappa^2 \left(\frac{\partial \mathbf{x}}{\partial \sigma} \right)^2 + \frac{1}{\omega} \left(\mathbf{\Pi} \frac{\partial \mathbf{x}}{\partial \sigma} \right)^2 + \kappa^2 \omega \right]^{1/2} \tag{4.17}$$

unless $\omega = 0$, and the Hamiltonian $\bar{H} = P_0 = \int_0^\pi H\, \mathrm{d}\sigma$ reproduces the equation of motion via $V_i = \delta \bar{H} / \delta \Pi_i$, $\partial \Pi_i / \partial t = -\delta \bar{H} / \delta x_i$, and there is no constraint.

Thus in the Hamiltonian formalism the main difference between our relativistic theory and the usual theory of a non-relativistic string becomes the occurrence of the square-root in the Hamiltonian density, and this is natural as relativization procedure. We also remark that in the case $\omega > 0$ we can consider the non-relativistic and weak-deformation limit, $|\mathbf{V}| \ll 0$ and $|\partial \mathbf{x}/\partial \sigma| \ll \sqrt{\omega}$. Then we have

$$\Pi_0 \approx \mu_0 + \frac{\mu_0}{2} \mathbf{V}^2 + \frac{K}{2} \mathbf{W}^2, \qquad \Pi_i \approx \Pi_0 V_i, \qquad S_i \approx K W_i.$$

These essentially coincide with the forms of non-relativistic string theory, and μ_0 and K represent the linear mass density and the modulus of tension.

Finally we remark that we can also expand $x_i(\sigma, t)$ and $\Pi_i(\sigma, t)$ in the form

$$x_i(\sigma, t) = \sum_{n=-\infty}^{\infty} x_i^n(t) \cos n\sigma,$$

$$\Pi_i(\sigma, t) = \frac{1}{2\pi} \sum_{n=-\infty}^{\infty} \Pi_i^n(t) \cos n\sigma,$$

$$(x_i^{-n} = x_i^n, \quad \Pi_i^{-n} = \Pi_i^n)$$

by virtue of the boundary conditions [See (4.6) and (4.12)].

Though it is difficult to obtain general solutions for the fundamental equations in this gauge, solutions with transversal internal movement are obtained directly in this gauge (See Section 7).

5. Geometric viewpoint

In the theory given in Sections 2–4 the case $\omega = 0$ is special because then the action (2.12) reduces to the area on the world strip

$$A = -\kappa \int_{\tau_0}^{\tau_1} d\tau \int_{\sigma_0}^{\sigma_1} d\sigma \sqrt{-D_0}, \tag{5.1}$$

which is invariant under the general transformation

$$\sigma \to \sigma'(\sigma, \tau), \qquad \tau \to \tau'(\sigma, \tau), \tag{5.2a}$$

$$\left.\frac{\partial \sigma'(\sigma, \tau)}{\partial \tau}\right|_{\sigma_0, \sigma_1} = 0, \tag{5.2b}$$ [10]

so that the equation of motion following therefrom is general-covariant.

However, this is exactly the model which is to be obtained from the geometric viewpoint. In this viewpoint the form of world strip and the motion of the end alone have physical meaning, and (σ, τ) are arbitrary Gauss coordinates specifying a point ('event') on the strip. Thus they have the arbitrariness of (5.2), where the new (σ', τ') give an equivalent representation of the same strip such that

$$x_\mu = x'_\mu(\sigma', \tau') = x_\mu(\sigma, \tau). \tag{5.3}$$

The geometrical quantities intrinsic to the strip are invariant under (5.2). Note that now both τ and σ need not be restricted to Lorentz scalars, and (5.2) may contain a transformation altering their dimensions. Any physically realizable strip must satisfy a certain equation of motion and boundary conditions, but these are physically meaningless unless they be covariant under (5.2). Thus they should be derived from an action integral which is invariant under (5.2). For the present we consider the free case. Then such an action is restricted to the unique form (5.1) [11, 12, 19].

The variational principle on (5.1) again yields (2.16), in which p_μ and S_μ are now

$$p_\mu = \kappa \frac{G^{11} x_{\mu,0} - G^{10} x_{\mu,1}}{\sqrt{-D_0}}, \qquad S_\mu = \kappa \frac{G^{10} x_{\mu,0} - G^{00} x_{\mu,1}}{\sqrt{-D_0}}, \tag{5.4}$$

and therefore we can bring (2.16) to the following form:

$$\eta_{\mu\nu} G_{\alpha\beta} \frac{\partial^2 x^\nu}{\partial\zeta_\alpha \partial\zeta_\beta} = 0, \qquad \left[G_{1\alpha} \frac{\partial x_\mu}{\partial\zeta_\alpha} \right]_{\sigma_0, \sigma_1} = 0. \tag{5.5a, b}$$

Here $(G_{\alpha\beta})$ is the inverse matrix of $(G^{\alpha\beta})$. More explicitly

$$G_{11} = G^{00}/D_0, \qquad G_{00} = G^{11}/D_0, \qquad G_{10} = G_{01} = -G^{10}/D_0.$$

$\eta_{\mu\nu}$ is a projector defined by

$$\eta_{\mu\nu} = \eta_{\nu\mu} = g_{\mu\nu} - G_{\alpha\beta} \frac{\partial x_\mu}{\partial\zeta_\alpha} \frac{\partial x_\nu}{\partial\zeta_\beta},$$

which satisfies $\eta_{\mu\nu}\eta^\nu_\lambda = \eta_{\mu\lambda}$, and

$$\eta_{\mu\nu} \frac{\partial x^\nu}{\partial\zeta_\alpha} = 0. \qquad (\alpha = 1, 0). \tag{5.6}$$

This model has only one structure constant κ and the theory is symmetric in σ and τ from the beginning, aside of the boundary condition (5.5b). The equation of motion must contain only *two* independent equations because the strip, which alone has physical sense, is a 2-dimensional surface. Indeed (5.5) contains just $4-2=2$ independent equations since it is covariant under the general transformation with respect to the two arbitrary parameters σ and τ. (In fact this is ensured by the indentity (5.6)).

Physically this means that each element of the string has only two degrees of freedom instead of three, because the longitudinal motion has no physical sense now. This is verified by noting that in the geometrical viewpoint observable quantities must be invariant under the general transformation (5.2). Now the 'velocity' V_i (except $V_i|_{0,\pi}$) and the 'radius vector' W_i defined by (2.7) and (2.10) are not invariant against (5.2) so that they are not observables in the geometrical model[11] but depend on the gauge. We can prove however that the 'normal velocity' $V_i^{\perp}$ orthogonal to the string

$$V_i^{\perp}=\zeta_{ik}V_k, \qquad (\zeta_{ik}\equiv\delta_{ik}-W_iW_k/\mathbf{W}^2, \qquad \zeta_{ik}W_k=0) \tag{5.7}$$

is invariant under (5.2) and is observable. The 4-momentum and the angular momentum tensor, P_μ and $M_{\mu\nu}$, which are given in the form (2.20) and (2.22) with the use of p_μ and S_μ of (5.4), are also invariant under (5.2). The relation (2.24) now splits into

$$p\frac{\partial x}{\partial\sigma}=0, \qquad p^2+\kappa^2\left(\frac{\partial x}{\partial\sigma}\right)^2=0. \tag{5.8}$$

The existence of these two local identities between p_μ, which is canonical conjugate to x_μ, and $\partial x/\partial\sigma$ reflects the covariance of theory under (5.2).

For further analysis of theory we choose a suitable gauge, and again there exist two important choices. The first one is to impose *two* local conditions in exploiting the fact that the theory is originally covariant under (5.2) which contains two arbitrary parameters. For such conditions it is natural to take

$$G^{10}=0, \qquad G^{00}+G^{11}=0, \tag{5.9a, b}$$

from the argument of Section 3. Then (5.5) simplify to the form (3.4), and both σ and τ are now Lorentz scalars. The two conditions (5.9a, b), however, are not mutually independent, so that the gauge is not yet fixed uniquely. Indeed the set of (5.9a, b) and (3.4a, b) is still invariant under the conformal transformation (3.10), *which now forms a subgroup of* (5.2). We are now fixing $[\sigma_0, \sigma_1]$ again as $[0, \pi]$. The conformal invariance means here that $x_\mu(\sigma,\tau)$ and $x_\mu^c(\sigma,\tau)=x_\mu(\sigma^c(\sigma,\tau), \tau^c(\sigma,\tau))$ trace out the same strip so that they represent one and the same motion of the string, even though they have different values of $C_\mu^r(r\neq 0)$. We note in passing that (5.9a, b) are represented also as

$$(v_\mu(\tau))^2 = 0, \tag{5.10}$$

and that, under (5.9a, b), the two light-like directions (2.3) on the strip become simply $\mathrm{d}\tau \pm \mathrm{d}\sigma = 0$. This means that $\tau \pm \sigma$ are really local light-like parameters.

To fix the gauge completely it is necessary to impose also the condition

$$n_\mu p^\mu = \text{const.}, \tag{5.11a}$$

from which follows also

$$n_\mu S^\mu = 0, \quad \text{i.e.} \quad n_\mu \left(G^{00} \frac{\partial x^\mu}{\partial \sigma} - G^{01} \frac{\partial x^\mu}{\partial \tau} \right) = 0, \tag{5.11b}$$

which is compatible with (5.5b). Here n_μ is a timelike (or light-like) constant vector. Under the constraint (5.9a), (5.11) above reduces to $n(\partial x/\partial \sigma) = 0$, which is conformal non-invariant of course [19]. We further get $nx(\sigma, \tau) = (\pi\kappa)^{-1}(nP)\tau + nX(0)$. As far as n_μ is an arbitrary constant vector independent of the motion, this gauge is a non-covariant one, but if we adopt P_μ for n_μ the gauge remains a covariant one.

To express the theory in the second choice of gauge we choose $\tau'(\sigma, \tau)$ such that it equals $x_0(\sigma, \tau)$. [See Equation (4.1).] Then, with the use of (4.2) we can prove that the equation of motion (5.5a) goes over to the non-covariant equation

$$\zeta_{ik} \left(\mathbf{W}^2 \frac{\partial V_k}{\partial t} - 2\mathbf{V}\mathbf{W} \frac{\partial V_k}{\partial \sigma} - (1 - \mathbf{V}^2) \frac{\partial W_k}{\partial \sigma} \right) = 0, \tag{5.12}$$

while the boundary condition is again (4.6)[12]. The invariant action integral (5.1) becomes

$$A = -\kappa \int_{t_0}^{t_1} \mathrm{d}t \int_0^\pi \sqrt{\Delta_0}\, \mathrm{d}\sigma, \tag{5.13}$$

where

$$\Delta_0 = \mathbf{W}^2(1 - \mathbf{V}^2) + (\mathbf{V}\mathbf{W})^2 = \mathbf{W}^2[1 - (\mathbf{V}^\perp)^2].$$

Thus

$$A = -\kappa \int_{t_0}^{t_1} \mathrm{d}t \int_0^{l_1} \sqrt{1 - (\mathbf{V}^\perp)^2} \cdot \mathrm{d}l, \tag{5.13'}$$

where l is the length given by (2.11), and the upper limit $l_1 = \int_0^\pi \sqrt{\mathbf{W}^2}\, d\sigma$ depends generally on t. The action (5.13), whence the equation of motion (5.12) which follows therefrom, are invariant under an arbitrary transformation

$$\sigma \to \tilde{\sigma}(\sigma). \tag{5.14}$$

If we keep the ends as $\tilde{\sigma}(0)=0$, $\tilde{\sigma}(\pi)=\pi$, (5.14) is expressed as

$$\tilde{\sigma}(\sigma) = \sigma + \sum_{n=1}^{\infty} b_n \sin n\sigma. \tag{5.14'}$$

Since (5.12) is invariant under (5.14) it contains two independent equations, as is obvious form $\zeta_{ik} \mathrm{W}_k = 0$, in agreement with the case of the original general covariant Equation (5.5a).

The energy and momentum densities per unit σ are now

$$\Pi_0(\sigma, t) = \kappa \mathbf{W}^2 / \sqrt{\Delta_0}, \qquad \Pi_i = \Pi_0 V_i^{\perp}, \tag{5.15a, b}$$

so that $V_i^{\parallel}$ is undetermined from Π_i and Π_0. Corresponding to (5.8) there hold the relations

$$\mathbf{\Pi} \frac{\partial \mathbf{x}}{\partial \sigma} = 0, \qquad \Pi_0^2 - \mathbf{\Pi}^2 = \kappa^2 \left(\frac{\partial \mathbf{x}}{\partial \sigma} \right)^2. \tag{5.16a, b}$$

Thus

$$\Pi_0 = \left[\mathbf{\Pi}^2 + \kappa^2 \left(\frac{\partial \mathbf{x}}{\partial \sigma} \right)^2 \right]^{1/2}. \tag{5.17}$$

The generator of the transformation (5.14) is $\mathbf{\Pi}(\sigma, t) \cdot \partial \mathbf{x}(\sigma, t)/\partial \sigma$, which however vanishes by (5.16a). Thus this symmetry does not supply physical conserved quantities. This implies that (5.14) is to be interpreted as a reparametrization. Thus in this case the choice (4.1) does not yet fix the gauge completely, corresponding to the fact that V_i and W_i still have arbitrariness depending on the gauge. Also Π_i and Π_0 depend on the gauge, but the momentum and energy densities per unit length of the string, q_μ, defined as

$$q_\mu = \Pi_\mu / \sqrt{\mathbf{W}^2}, \qquad \left(P_\mu = \int_0^{l_1} q_\mu \, dl \right) \tag{5.18}$$

are independent of the gauge, and are expressed as

$$q_0 = \kappa/\sqrt{1-(\mathbf{V}^\perp)^2}, \qquad q_i = q_0 V_i^\perp, \tag{5.19}$$

satisfying $q_0^2 - \mathbf{q}^2 = \kappa^2$, $\mathbf{qW} = 0$. To fix the gauge completely, we need to impose a constraint, which is (5.11). If we take P_μ for n_μ, this means

$$(P_0 - \mathbf{PV})(\mathbf{VW}) = \mathbf{PW}(1 - \mathbf{V}^2), \tag{5.20a}$$

$$(P_0 - \mathbf{PV})^2 \mathbf{W}^2 = [(m^2/\pi\kappa)^2 - (\mathbf{PW})^2](1 - \mathbf{V}^2). \tag{5.20b}$$

The gauge is now completely fixed and V_i and W_i become unambiguous. We see in particular that $\mathbf{V}$ and $\mathbf{V}^\perp$ are the same at the ends and that $\mathbf{V}^2 = 1$ (velocity of light) there. Also we see that in the center-of-mass rest frame ($\mathbf{P} = 0$) $\mathbf{VW} = 0$ everywhere. The present gauge means therefore the one in which in the CM frame $\mathbf{V}^\parallel$ vanishes and $\mathbf{V}$ equals the observable quantity $\mathbf{V}^\perp$ everywhere. Clearly, the nonrelativistic limit does not exist for the geometrical model because the ends move with the velocity of light.

6. Transversality Condition

We now return to the 'realistic' theory. This theory, as described in Sections 2–4, is complete in itself but we note that this theory still allows the introduction of a certain *physical* constraint consistently. The interest in this constraint comes from the fact that it produces a particular simplification of the theory and also that with its introduction the theory becomes common both for $\omega \neq 0$ and for $\omega = 0$ cases.

The constraint in question is

$$P_\mu p^\mu = \text{const.} \tag{6.1a}$$

This results in

$$P_\mu S^\mu = 0, \quad \text{i.e.} \quad P_\mu \left(G^{00} \frac{\partial x^\mu}{\partial \sigma} - G^{01} \frac{\partial x^\mu}{\partial \tau} \right) = 0, \tag{6.1b}$$

via (2.16a, b). Physically the new constraint restricts the internal motion to transversal ones (rotation, lateral vibration) alone, and thus reduces the degree of freedom of each 'parton' of the string from three to two.

If we first consider in the covariant gauge stated in Section 3, (6.1b) simplifies to

$$P\frac{\partial x}{\partial \sigma}=0. \tag{6.2}$$

This yields

$$Px(\sigma,\tau)=\frac{P^2}{\pi\kappa}\tau+PX(0). \tag{6.3}$$

In terms of the normal mode amplitude, (6.2) is expressed as

$$P^\mu C^r_\mu=0, \qquad (r\neq 0) \tag{6.4}$$

meaning that in the CM-rest frame all normal modes can only have space components ($C^r_0=0$).

The physical meaning of the new constraint becomes clearer by considering in the non-covariant formalism given in Section 4. First we note that in the non-covariant gauge, (6.1) takes the form (5.20) with the replacement $\mathbf{W}^2\to\mathbf{W}^2+\omega$ in the left side of (5.20b). Henceforth in this and the next sections we always consider in the CM rest frame where $\mathbf{P}=0$, $P_0^2=m^2$. Then (6.2) and (6.3) mean

$$\mathbf{VW}=0, \tag{6.5}$$

$$\mathbf{V}^2+v^2\mathbf{W}^2=b^2. \tag{6.6}$$

$$\left(v=\frac{\pi\kappa}{m},\quad b=\left[1-\frac{\pi^2\mu_0^2}{m^2}\right]^{1/2}\leqq 1\right).$$

We have also $\Delta_\omega=(1-\mathbf{V}^2)^2/v^2$. Then the non-linear equation of motion (4.5) simplifies to the *linear* equation

$$\frac{\partial^2 x_i}{\partial t^2}-v^2\frac{\partial^2 x_i}{\partial\sigma^2}=0. \tag{6.7}$$

This is formally analogous to the equation of motion of the nonrelativistic string, but the physical ingredient is quite different. We have otherwise the boundary condition.

$$W_i|_{0,\pi}=0. \tag{6.8}$$

The energy and momentum densities per unit σ become

$$\Pi_0(\sigma,t)=P_0/\pi=p_0=\text{const}, \qquad \Pi_i=\Pi_0 V_i=p_i, \tag{6.9a, b}$$

so that we need not distinguish Π_μ from p_μ, and Π_i agrees with V_i apart from the constant factor Π_0. Accordingly the rest frame condition is expressed as

$$\int_0^\pi V_i(\sigma, t)\,\mathrm{d}\sigma = 0. \tag{6.10}$$

Moreover the mechanical center-of-mass (4.15) becomes

$$\tilde{X}_i(t) = \frac{1}{\pi}\int_0^\pi x_i(\sigma, t)\,\mathrm{d}\sigma = \tilde{X}_i(0),$$

and this coincides with the geometrical center-of-mass coordinate $X_i(\tau) = X_i(0)$, so that we need not distinguish them. As (6.5) indicates, each portion of the string always moves perpendicular to the string, and it is important that this is valid irrespective of whether $\omega > 0$ or $\omega = 0$.

We insert here a note about the case $\omega = 0$, which is the case of the geometric model of Section 5. Here the equation of motion is (5.12), which gives only

$$\zeta_{ik}(\partial V_k/\partial t - v^2 \partial W_k/\partial\sigma) = 0 \tag{6.11}$$

in the rest frame. On the other hand we now have the constraints (6.5) and (6.6) also, i.e. $\mathbf{VW} = 0$, $\mathbf{V}^2 + v^2\mathbf{W}^2 = 1$, which lead to $W_k(\partial V_k/\partial t - v^2 \partial W_k/\partial\sigma) = 0$. This relation and (6.11) result in (6.7). Thus under the present constraint the basic equations are common both to $\omega > 0$ and to $\omega = 0$ cases. As stated in Section 5, in the $\omega = 0$ case the constraint (6.2) is one we introduced to fix the gauge conveniently (rather than a physical constraint), and in this gauge $\mathbf{V}$ and $\mathbf{V}^\perp$ are the same in the CM rest frame.

According to (6.6) the velocity is largest at both ends where it is $|\mathbf{V}| = b$, which approaches the light velocity with $m \to \infty$ if $\omega > 0$, and is always c if $\omega = 0$. Equation (6.6) is rewritten as

$$\Pi_0 = \left[\frac{\mathbf{W}^2 + \omega}{1 - \mathbf{V}^2}\right]^{1/2} = \text{const} = \frac{m}{\pi} \tag{6.12}$$

Equations (6.9b) and (6.12) imply that the *local* rest mass density per unit σ is

$$\mu = \kappa\sqrt{\mathbf{W}^2 + \omega} = \sqrt{\mu_0^2 + \kappa^2\mathbf{W}^2}. \tag{6.13}$$

7. Analysis of the Motion, and Solutions

We consider the properties of the solutions to the fundamental Equations (6.5)–(6.8) and (6.10), which are common to the 'realistic model' under the transversality condition and to the 'geometric model'. Thus the following analysis apply to both cases.

To solve the fundamental equations we use the fact that the whole motion is again determined once we are given the motion of the end, i.e. $x_i(0, t)$ and $\partial x_i(0, t)/\partial t = V_i(0, t) \equiv V_i^0(t)$, where $V_i^0(t)$ must satisfy the following three conditions:

$$V_i^0(t+2\pi/\nu) = V_i^0(t), \tag{7.1}$$

$$(\mathbf{V}^0(t))^2 = b^2 \tag{7.2}$$

$$\int_0^{2\pi/\nu} V_i^0(t)\,\mathrm{d}t = 0. \tag{7.3}$$

The solution is then given by

$$\begin{aligned} x_i(\sigma, t) &= \tfrac{1}{2}\{x_i(0, t-\sigma/\nu) + x_i(0, t+\sigma/\nu)\}, \\ V_i(\sigma, t) &= \tfrac{1}{2}\{V_i^0(t-\sigma/\nu) + V_i^0(t+\sigma/\nu)\}. \end{aligned} \tag{7.4}$$

Thus the motion has the following remarkable features in the rest frame.

(i) Since the motion of each end is periodic (See (7.1)) and transversal and moreover *has constant speed*, it must be a *rotational* motion.

(ii) On account of the condition (7.3) i.e. (6.10), not only $V_i(\sigma, t)$ but $x_i(\sigma, t)$ itself is periodic in t, $x_i(\sigma, t+2\pi/\nu) = x_i(\sigma, t)$, with the period

$$\Delta t = 2\pi/\nu = 2m/\kappa. \tag{7.5}$$

This means that each portion of the string moves in the CM-frame along a *closed orbit* with the fundamental period (7.5). This is somewhat reminiscent of a quantized motion in the old quantum theory, and the condition (7.3) simulates the quantum condition. This periodic motion of each portion, other than the ends, is generally rotational and vibrational.

(iii) The basic frequency $\nu/2\pi = \kappa/2m$ is *inversely* proportional to the mass. This fact might appear to be opposite to de Broglie's relation [8] $mc^2 = h\nu$, which was basis of de Broglie's discovery of wave mechanics, but the above fact is related to causality. A higher mass state has a larger mean size and in order that in its rotation the velocity of the ends does not exceed c the angular velocity must be smaller.

Now the function $V_i^0(t)$ satisfying the above conditions is written in the form

$$V_1^0(t)+iV_2^0(t)=b\sin\theta(T)\cdot e^{i\varphi(T)},\qquad V_3^0(t)=b\cos\theta(T), \tag{7.6}$$

with $T=vt$ and $\theta(T)=n_\theta T+g(T)$, $\varphi(T)=n_\varphi T+h(T)$, where n_θ and n_φ are integers while $g(T)$ and $h(T)$ are periodic functions, $g(T+2\pi)=g(T)$, etc. They are not completely arbitrary, since we must have

$$\int_0^{2\pi}\cos\theta(T)\,\mathrm{d}T=0,\qquad \int_0^{2\pi}\sin\theta(T)\,e^{i\varphi(T)}\,\mathrm{d}T=0, \tag{7.7}$$

to satisfy (7.3).

We now illustrate the motion of the string by some typical solutions.

(i) First we consider the case where only the r-th mode is excited. In this case the transversality condition is automatically satisfied. This solution is given by taking $\theta(T)=\pi/2$ and $\varphi(T)=rT+\text{const}$ in (7.6). The motion is given by

$$\begin{pmatrix}x_1\\x_2\end{pmatrix}=\frac{b}{rv}\cos r\sigma\begin{pmatrix}\sin rvt\\\cos rvt\end{pmatrix},\qquad (x_3=0). \tag{7.8}$$

This has r nodes, at $\sigma=(n+\frac{1}{2})\pi/r$ $(n=0,1,\ldots,r-1)$, where $\mathbf{V}=0$, whereas $\mathbf{W}$ vanishes at $\sigma=n\pi/r$ $(n=0,1,\ldots,r)$. The elementary length along the string (in the CM-rest frame) is $\mathrm{d}l=\sqrt{\mathbf{W}^2}\cdot\mathrm{d}\sigma=(b/v)\sin r\sigma\cdot\mathrm{d}\sigma$.

(a) If $r=1$, the string is a straight line which rotates rigidly around its center $\sigma=\pi/2$ with the constant angular velocity v, and with the constant length $l_1=l(\pi)=2b/v$. This solution is the one already given in the covariant formalism (Section 3) and constitutes the leading trajectory (3.23).

(b) If $r=2$, the string is folded in such a way that the points σ and $\pi-\sigma$ coincide, and it rotates around the point $\sigma=\pi/4$ (i.e. $\sigma=3\pi/4$) with the angular velocity $2v$.

(c) Similarly, for $r=r$, the string is r-ply folded and makes a pure rotation with the angular velocity rv. Thus we find that our relativistic string implies a striking realization of ‘relativistic rotator model’. The magnitude of spin is

$$J = m^2 b^2/(2\pi\kappa r) = (m^2 - \pi^2\mu_0^2)/(2\pi\kappa r),$$

while the actual length of the string is $\bar{l}_r = \int_0^{\pi/r} \sqrt{\mathbf{W}^2}\,\mathrm{d}\sigma = 2\sqrt{2J/(\pi\kappa r)}$, so that

$$\frac{m\bar{l}_r}{J} = 4\left(1 - \frac{\pi^2\mu_0^2}{m^2}\right)^{-1/2} \geqq 4. \tag{7.9}$$

This relation verifies that Møller's inequality [18] $\bar{l} \geqq J/m$ between spin, mass and the spatial extension $\bar{l}$ for any system extended with positive energy density is fulfilled for our motion, as it should.

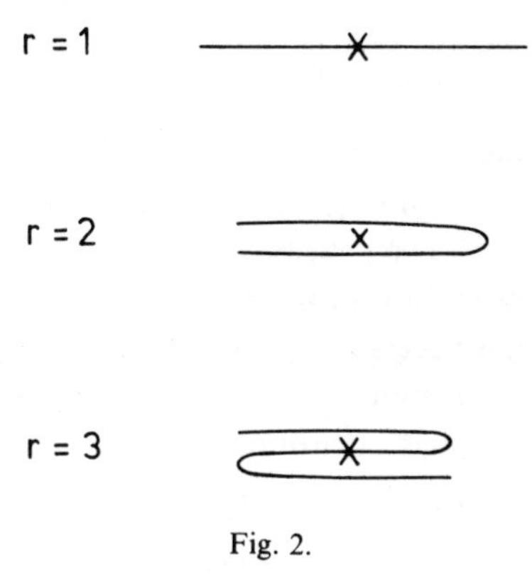

Fig. 2.

(ii) Another typical example of a plane motion ($\theta(T) = \pi/2$) is given by $\varphi(T) = \alpha \sin T$, ($\alpha$ = const). Then

$$V_1^0 + iV_2^0 = b\, e^{i\alpha \sin T} = b \sum_{m=-\infty}^{\infty} J_m(\alpha)\, e^{imvt}.$$

The motion is the superposition of vibrations with angular frequencies of α to the roots of Bessel function: $J_0(\alpha) = 0$, i.e. $\alpha = \lambda_k^{(0)}$, $(k = 1, 2, \ldots)$. This implies the condition of closed orbit. The motion involves all higher harmonic vibrations, and the spin is found to be $J = 0$.

(iii) A simple example of a non-planar motion is given by taking

$$\theta(T) = T + \alpha, \quad \varphi(T) = 2T + \beta, \qquad (\alpha, \beta = \text{const}).$$

The condition (7.7), i.e. $\int_0^{2\pi} e^{i\alpha \sin T}\,\mathrm{d}T = 0$, 'quantizes' the admissible values

ν and 3ν. The string executes 3-dimensional vibrations and rotations around the nodal point $\sigma=\pi/2$.

All the solutions given above are those satisfying the transversality condition. In the case of the 'realistic' model ($\omega\neq 0$), there are of course many more solutions originally which involve longitudinal movement as well. For obtaining those solutions the Lorentz covariant formalism (Section 3) is more convenient. An example of such solution was explicitly given in Ref. [15].

We have seen that the transversal solutions have remarkably simple properties, but it is to be noted that in order to derive the dual amplitude in quantum theory the whole set of solutions must be taken into account.

8. QUANTUM THEORY

Since we have clarified the structure of the classical theory of the free relativistic string we can now proceed to quantum theory and also to interactions, based on it. The process is not straightforward, but we limit ourselves to a very brief description here.

For quantization it is necessary to consider in a certain fixed gauge. The most convenient is the Lorentz covariant gauge (stated in Section 3), in which we now put the equal τ commutation relation

$$[x_\mu(\sigma,\tau),\, p_\nu(\sigma',\tau)]=i\hbar g_{\mu\nu}[\delta(\sigma-\sigma')+\delta(\sigma+\sigma')+ +\delta(2\pi-\sigma-\sigma')], \tag{8.1}$$

in the Heisenberg picture regarding τ. We then obtain, in free case, the more general commutation relation [9]

$$[x_\mu(\sigma,\tau),\, x_\nu(\sigma',\tau')]=-i(\hbar/\kappa)\, g_{\mu\nu}\Delta(\sigma,\sigma',\tau-\tau'). \tag{8.2}$$

This implies that

$$[x_\mu(\sigma,\tau),\, x_\nu(\sigma',\tau')]=0,\quad \text{for}\quad |\tau-\tau'|<|\sigma-\sigma'|, \tag{8.3}$$

expressing causality when $G^{11}\geqq 0$ is guaranteed. Also from (8.2) $u_\mu(\tau)= =x_\mu(0,\tau)$ and $v_\mu(\tau)=\partial x_\mu(0,\tau)/\partial\tau$ satisfy the commutation relation [17]

$$[u_\mu(\tau),\, v_\nu(\tau')]=(2i\hbar/\kappa)\, g_{\mu\nu}\sum_n \delta(\tau-\tau'-2n\pi). \tag{8.4}$$

It is proved that the above quantization is invariant under conformal transformation (3.10), and Equation (8.4) implies that $v(\tau)$ has 'conformal spin' 1. The conformal operators Λ^r have been given in Section 3, but in particular Λ^0 (the scalar Hamiltonian) must now be defined by the normal product $\Lambda^0 = \pi\kappa \sum_n : C^n C^{-n} : = P^2/(2\pi\kappa) + 2\pi\kappa \sum_{n=1}^{\infty} Cn^\dagger C^n$. Corresponding to the classical constraints (3.17) the state vector must satisfy the wave equation and subsidiary conditions

$$\left(\Lambda^0 + \frac{\pi\kappa\omega}{2}\right)\Psi = 0, \qquad \Lambda^r \Psi = 0, \qquad (r = 1, 2, \ldots). \qquad (8.5\text{a, b})$$

The conformal operators lose their closure property as a whole, such that $[\Lambda^r, \Lambda^s] = \hbar(r-s)\,\Lambda^{r+s} + \hbar^2 \delta_{r+s,\,0}\, r(r^2-1)/3$ [9]. On account of this fact and also of the fact that ω is generally non-zero, the compatibility requires that we should restrict the subsidiary conditions (8.5) to the range of $r = 0, 1, 2, \ldots.$ alone, as indicated. In classical theory the case $\omega = 0$ had the special feature that the whole theory is conformal-invariant in this case only, whereas in quantum mechanics the whole theory is invariant under the nonunitary transformations generated by $\Lambda^r + (\pi\kappa\omega/2)\,\delta_{r0}\,(r = 0, 1, 2, \ldots)$ regardless of whether $\omega \neq 0$ or $\omega = 0$, and thus the $\omega = 0$ case loses its special position. When we assume (8.5) we are also assuming an indefinite metric with respect to relative-time degrees, and (8.5) determines the mass spectrum. In the Schrödinger picture regarding τ, τ is completely eliminated [15], so that we call this representation as 'σ-formalism' [17], where the theory is represented by (8.5), or equivalently by the super-wave equation defined at every point of the string [9]

$$\left[\tfrac{1}{2}:\frac{(p_\mu)^2}{\kappa} + \kappa\left(\frac{\partial x_\mu}{\partial \sigma}\right)^2 : + \frac{i}{2\pi}\int_0^\pi \left\{p_\mu(\sigma'),\right.\right.$$

$$\left.\left.\frac{\partial x^\mu(\sigma')}{\partial \sigma'}\right\} \frac{\sin\sigma'\,\mathrm{d}\sigma'}{\cos\sigma - \cos\sigma'} + \frac{\kappa\omega}{2}\right]\Psi = 0.$$

It is expected for correspondence-theoretical reasons that the possible negative-norm states are suppressed precisely by (8.5). This was recently verified [20] for $\pi\kappa\omega/2 = -\hbar$.

Due to the 'symmetry' in σ and τ, we should be able to represent the theory also in the 'τ-formalism' in which σ disappears [17]. This means that we can also represent the theory of the relativistic string in terms of the motion of the end $u_\mu(\tau)$ and $v_\mu(\tau)$, in quantum theory as well. This fact is easily seen, based on the commutation relation (8.4) and the expression (See (3.15))

$$\Lambda^r = \frac{\kappa}{4} \int_{-\pi}^{\pi} :(v_\mu(\tau))^2: e^{ir\tau}\, \mathrm{d}\tau. \tag{8.6}$$

We remark in passing that the transversality condition stated in Section 6 can be introduced in quantum theory. For this purpose we impose the additional subsidiary condition

$$P_\mu \left(\frac{\partial x^\mu}{\partial \sigma} \right)^{(+)} \Psi = 0, \tag{8.7}$$

corresponding to the classical constraint (6.2). Here we again extracted the positive-frequency part $(\partial x_\mu/\partial\sigma)^{(+)}$ from $\partial x_\mu/\partial\sigma$, and it is clear that (8.7) is reexpressed as

$$P^\mu C^r_\mu \Psi = 0, \qquad (r = 1, 2, \ldots). \tag{8.8}$$

The quantization of the classical theory of the relativistic string can also be performed on the basis of the non-covariant formalism given in Section 4. In the Schrödinger picture regarding t the quantum mechanics is given by

$$\begin{gathered} [x_i(\sigma), \Pi_j(\sigma')] = i\hbar \delta_{ij} (\delta(\sigma - \sigma') + \delta(\sigma + \sigma') + \delta(2\pi - \sigma - \sigma')), \\ i\hbar\, \partial\Phi/\partial t = \bar{H}\Phi, \qquad (\Phi = \Phi[\mathbf{x}(\sigma), t]) \end{gathered} \tag{8.9}$$

where $\bar{H} = \int_0^\pi H\, \mathrm{d}\sigma$ is to be given from (4.17) for the case $\omega > 0$, but we now need to take the normal product:

$$H(\sigma) = \left[:\mathbf{\Pi}^2 + \kappa^2 \left(\frac{\partial \mathbf{x}}{\partial \sigma} \right)^2 : + \frac{1}{4\omega} :\left\{ \mathbf{\Pi} \frac{\partial \mathbf{x}}{\partial \sigma} \right\}^2 : + \kappa^2 \omega \right]^{1/2}. \tag{8.10}$$

In the non-covariant formalism the quantity $x_0(\sigma)$ is completely eliminated so that we are entirely free from indefinite metric, but in the form above given it looks to be mathematically difficult to solve the equations leading to mass spectrum. Moreover in the non-covariant quantum theory problem arises in the proof of Lorentz invariance, due to the redefinition by normal product. Nevertheless, if we now introduce the transversality constraint the theory is simplified and can be solved.

To introduce the coupling with an external scalar field $\phi(x)$ we again start from the Lorentz covariant gauge and add the interaction term acting at the end [9]

$$L_1' = -\frac{2g}{\pi}\,\delta(\sigma)\,\phi(x(0,\tau)) \tag{8.11}$$

to the free Lagrangian L_1 of (3.6).

We assume a plane wave $\phi(x)=e^{ikx}\,\phi(0)$ with $-k^2=m_{ex}^2/\hbar^2$ and understand the normal product. The wave equation and subsidiary conditions (8.5) are modified to

$$\left[\Lambda^r + \frac{g}{\pi}{:}\phi(x(0)){:} + \frac{\pi\kappa\omega}{2}\,\delta_{r0}\right]\Psi = 0, \qquad (r=0, 1, 2, \ldots) \tag{8.12}$$

which are mutually compatible only when the external mass has the unphysical Virasoro value $m_{ex}^2 = -2\pi\kappa\hbar$ [21]. If we further identify the external scalar meson with the ground mass state of the string, we have again $\pi\kappa\omega/2 = -\hbar$, which means that the leading intercept is $\alpha(0)=1$. It is well-known that in this case the quantum mechanical transition amplitude obtained from the propagator and the interaction vertex implied in (8.12) reproduce the dual amplitude [5]. We refrain, however, from entering into closer discussion of interaction in this paper.

9. Concluding Remarks

To conclude we want to stress a few general points.

(i) Our string theory for hadrons implies a genuine spacetime model. In our theory it is essential that the basic variables $x_\mu(\sigma, \tau)$ have the meaning of space-time coordinates, because it is on the basis of this interpretation that we can properly identify the 4-momentum P_μ of the system [9]. (See Equations (2.20), (4.7), (8.1)), formulate causality (See Equations (2.8), (8.3)), and locate interactions (See Equation (8.11)).

(ii) The variables x_μ play the dual role: They themselves physically represent the 4-dimensional positional coordinates, but at the same time they are viewed like a field in the 2-dimensional space of Gauss coordinates (σ, τ). Thus our string theory which is still at the one-particle level is formally analogous to a field theory in 2-dimensional space-time, where the characteristic 'symmetry' in σ and τ is at work. Moreover the 'general covariance' under (5.2) or the 'partial covariance' under (2.5) on (σ, τ) allows the string theory be analogous to the general theory of relativity or to the gauge theory. However, there arise simple and peculiar aspects of the string theory due to the 2-dimensional nature of the parameter space. Thus the equation of motion, which is originally highly nonlinear, can be linearized in a suitable gauge. The existence of conformal symmetry is another such aspect.

(iii) The 'realistic' viewpoint has led us to the theory of relativistic string which is more general than the theory obtained from the 'geometric' viewpoint and includes the latter theory as a singular limiting case $(\omega = 0)$ where some special features arise in the classical level by higher symmetry. A drawback of the geometrical viewpoint is that in quantization and interaction one is obliged to amend the theory arbitrarily by introducing a non-zero ω. (cf. Ref. [19]). Another drawback of the geometrical approach is that it does not lead to a suitable classical foundation for the quantum theory in the presence of coupling with external scalar field.

(iv) The pure string theory treated in this paper is only an ideal one for a model of hadrons. One defect is that it does not contain isospin, half-integer spin and baryon number properties. The generalization towards the introduction of these properties will become possible by incorporating some features of the relativistic rotator model and also the fusion theory initiated by de Broglie [22]. The pure string model can be regarded as the limit of a linear chain consisting of mass points, but now we consider a string corresponding to the limit of linear chain consisting of Dirac-like elements, and then the theory is linearized. The result represents baryons or mesons depending on whether we linearize both the wave equation and subsidiary condition or the subsidiary conditions alone (cf. Refs. [23, 4, 17]).

This work is a development of my talk at the Xth Coral Gables Conferences (January 1973) and my lecture at the Institut Henri Poincaré (February 1973). I would like to thank Professors B. Kursunoglu, B. Sakita, K. Kikkawa, Y. Nambu, and J.-P. Vigier for valuable discussions at various places on my winter journey.

Nagoya University

NOTES

* Paper dedicated to the Prof. L. de Broglie anniversary.

[1] It is also interesting to recall that in the remote past a string suggested harmony at the birth of theoretical physics thinking by Pythagorus and that later Kepler compared the revolutions of planets around the sun to the vibrations of a string.

[2] We employ the notation convention such as $(\mathrm{d}x_\mu)^2 = \mathrm{d}x_\mu\, \mathrm{d}x^\mu = \sum_i (\mathrm{d}x_i)^2 - (\mathrm{d}x_0)^2$.

[3] In the special case $\omega = 0$, (2.16a) becomes, instead (5.5a).

[4] To construct the theory of relativistic string it is possible directly to start from (3.6), i.e. (3.4), which are not invariant under (2.5) but are invariant under conformal transformation. In this method we first note that (3.4a, b) are not sufficient to define a physical model because they do not ensure the 'uniqueness of physical interpretation' nor causality. It is then proved [15] that to ensure these requirements we must impose the condition (3.5), which works to break the conformal symmetry.

[5] The latter relation is necessary to leave (3.4b) invariant.

[6] Strictly speaking we have $\partial x_0/\partial \tau = \pm\sqrt{-G^{00}}/\sqrt{1-\mathbf{V}^2}$, but because $\partial x_0/\partial \tau$ has a definite sign (either positive or negative) everywhere we have written as in (3.19), taking the positive case.

[7] If $\omega = 0$ the variational principle yields Equation (5.12).

[8] The suffix μ in Π_μ as well as in $\tilde{S}_\mu$ (See Equation (4.9)) is generally not a 4-vector index. Note however that *at each end* Π_μ is a 4-vector such that $\Pi_\mu = \mu_0 U_\mu$ and coincides with p_μ in the Lorentz covariant gauge. Also $\tilde{S}_\mu = 0$ at the ends.

[9] Again μ, ν in $\mathfrak{M}_{\mu\nu}$ are generally not 4-vector indices.

[10] The boundary condition (5.2b) is necessary in order that the boundaries map the boundaries. The transformations $\sigma \to \tau$, $\tau \to \sigma$, and $\sigma \to \sigma' = \tau - \sigma$, $\tau \to \tau' = \tau + \sigma$ are quite useful but do not belong to (5.2).

[11] $V_i\,|_{0,\pi}$ are invariant against (5.2a) due to (5.2b).

[12] The condition $W_i\,|_{0,\pi} = 0$ is invariant against (5.2) although the quantities $W_i\,|_{0,\pi}$ are not.

BIBLIOGRAPHY

[1] De Broglie, L., Bohm, D., Hillion, P., Halbwachs, F., Takabayasi, T., and Vigier, J. P., *Phys. Rev.* **129** (1963), 438.

[2] See also Bohm, D. and Vigier, J. P., *Phys. Rev.* **109** (1958), 1882; Takabayasi, T., *Prog. Theor. Phys.* **23** (1960), 915; *Prog. Theor. Phys. Suppl.* Extra Number (1965), 339.
Hara, O. and Goto, T., *Prog. Theor. Phys. Suppl.* No. 41 (1968), 56.

[3] Takabayasi, T., *Prog. Theor. Phys.* **34** (1965), 124; *Prog. Theor. Phys. Suppl.* No. 41 (1968), 130 (See pp. 159–160 thereof).
Fujimura, K., Kobayashi, T., and Namiki, M., *Prog. Theor. Phys.* **43** (1970), 73.

[4] Takabayasi, T., *Prog. Theor. Phys.* **48** (1972), 1718.

[5] Nambu, Y., *Symmetries and Quark Models*, Gordon and Breach, New York, 1970, p. 269. Susskind, L., *Nuovo Cimento* **69A** (1970), 457.

[6] Takabayasi, T., *Prog. Theor. Phys.* **43** (1970), 1117.

[7] Veneziano, G., *Nuovo Cimento* **57A** (1968), 190.

[8] de Broglie, L., *Comptes Rendus* **177** (1923), 507, 548; Thèse (1924).

[9] Takabayasi, T., *Prog. Theor. Phys.* **44** (1970), 1429; **46** (1971), 1528 and 1924.

[10] Hara, O., *Prog. Theor. Phys.* **46** (1971), 1549.

[11] Goto, T., *Prog. Theor. Phys.* **46** (1971), 1560.
[12] Mansouri, F. and Nambu, Y., *Phys. Letters* **39B** (1972), 375.
[13] Chang, L. N. and Mansouri, J., *Phys. Rev.* **D5** (1972), 2535.
[14] Konisi, G., *Prog. Theor. Phys.* **48** (1972), 2008.
[15] Takabayasi, T., *Prog. Theor. Phys.* **49** (1973), 1724; Lecture at Coral Gables Conferences on Fundamental Interactions (January 1973).
[16] de Broglie, L., *Une nouvelle conception de la Lumière*, Hermann, Paris, 1934. Proca, A., *Journ. Phys. Rad.* **7** (1936), 347. Bass, L. and Schrödinger, E., *Proc. Roy. Soc.* **A232** (1955), 1.
[17] Takabayasi, T., *Prog. Theor. Phys.* **47** (1972), 1026.
[18] Møller, C., *Ann. Inst. H. Poincaré* **11** (1949), 251.
[19] Goddard, P., Goldstone, J., Rebbi, C., and Thorn, C. B., CERN preprint.
[20] Brower, R., *Phys. Rev.* **D6** (1972), 1655.
[21] Virasoro, M. A., *Phys. Rev.* **D1** (1970), 2933.
[22] de Broglie, L., *Théorie générale des particules à spin*, Gauthier-Villars, Paris, 1943.
[23] Ramond, P., *Phys. Rev.* **D3** (1971), 2415. Neveu, A. and Schwarz, J. H., *Nuclear Phys.* **B31** (1971), 86.

Note added in proof: During these two years since this paper was written many papers have appeared on some aspects of the theory of relativistic string. Here I only mention Takabayasi, T., *Prog. Theor. Phys.* **52** (1974), 1910, which supplements the treatment of 'geometric' string in the present paper.

J. GÉHÉNIAU

ON THE PHOTON THEORY OF L. DE BROGLIE

The photon theory initiated by L. de Broglie some forty years ago contains several characteristic features, the most important of which I shall review in the following pages: a representation of the photon wave function through a bispinor, its wave equation, the annihilated state, and three points which still raise interesting questions, i.e. complex electromagnetic fields for elementary phenomena, non-Maxwellian tensor densities, and a massive photon.

1. The Wave Function

Soon after the publication of Dirac's relativistic electron theory, L. de Broglie proposed to consider the photon as a point system formed by 'fusion' of a pair of spin 1/2 'complementary' particles; the wave function of this particle was thus a bispinor [1 (a), (1934), p. 38 and (b) Vol. I, p. 146].

Its sixteen components in [1 (a) and (b)] are those which transform like the product of a spinor by its complex conjugate. Here we shall use mainly, as does Mme Tonnelat in her thesis [2], the ϕ_i^k which transform like the products of the components ψ_i of a spinor ψ by those, $\bar{\psi}^k$, of its adjoint $\bar{\psi}=\psi^+\gamma_0$ where $\psi^+={}^t\psi^*$ is the transported conjugate of ψ, and γ_0 is the first of the Dirac matrices γ_α:

$$\gamma_\alpha\gamma_\beta+\gamma_\beta\gamma_\alpha=2\eta_{\alpha\beta}I \qquad (\alpha, \beta=0, 1, 2, 3)$$

$$\eta_{00}=1, \qquad \eta_{11}=\eta_{22}=\eta_{33}=-1$$

$$\eta_{\alpha\beta}=0 \quad \text{if} \quad \alpha\neq\beta.$$

The ϕ_i^k are the elements of the matrix ϕ; i indicates the lines, k the columns. Through the known reduction formula

$$4\phi=SI+A_\alpha\gamma^\alpha+\tfrac{1}{2}H_{\alpha\beta}\sigma^{\alpha\beta}+B_\alpha\gamma_5\gamma^\alpha+Pi\gamma_5, \tag{1}$$

where

$$\sigma_{\alpha\beta}=i\gamma_\alpha\gamma_\beta, \qquad \gamma_5=i\gamma^0\gamma^1\gamma^2\gamma^3, \qquad (\alpha\neq\beta)$$

each ϕ_i^k is given as a linear combination of the components of a scalar S,

M. Flato et al. (eds.), Quantum Mechanics, Determinism, Causality, and Particles, 217–225. *All Rights Reserved. Copyright © 1976 by D. Reidel Publishing Company, Dordrecht-Holland.*

a vector A_α, an antisymmetrical tensor $H_{\alpha\beta}$, a pseudo-vector B_α and a pseudo-scalar P, and conversely these tensors are linear combinations of the ϕ_i^k; for instance,

$$A_\alpha = \operatorname{tr} \gamma_\alpha \phi \tag{2}$$

where tr = trace of.

We note ϕ^+ the transposed conjugate of ϕ and call adjoint of ϕ

$$\bar{\phi} = \gamma_0 \phi^+ \gamma_0. \tag{3}$$

This bispinor transforms like ϕ and its reduction formula is just (1) but with the conjugates S^*, ...instead of S, ... The reality condition of those tensors is thus

$$\phi = \bar{\phi} \tag{4}$$

Let us remark that, with the matrix of 'charge' conjugation C,

$$-C \cdot {}^t\gamma_\alpha = \gamma_\alpha C, \tag{5}$$

one gets, [1, (d), Ch. VII], a

$$\Psi = \phi C \tag{6}$$

whose components transform like the products of two spinors. The tensors A_α, $H_{\alpha\beta}$ are built with the symmetrical part

$$\Psi_{ik}^{(1)} = (\Psi_{ik} + \Psi_{ki})/2 \tag{7}$$

the tensors S, B_α, P with the antisymmetrical part of Ψ.

2. The Wave Equation

We shall first write the wave equation for Ψ, and note $\gamma_{(1)}^\alpha$, $(\gamma_{(2)}^\alpha)$. Dirac matrices which operate on the first (second) index of Ψ_{ik}. The two complementary particles appear symmetrically in ψ_{ik}; so the $\gamma_{(1)}^\alpha$ and $\gamma_{(2)}^\alpha$ are a couple of identical matrices γ^α.

With these notations the de Broglie wave equation for the photon is written

$$p\Gamma\Psi(x) = \kappa\Psi(x). \tag{8}$$

where

$$p_\alpha = i\frac{\partial}{\partial x^\alpha} = i\partial_\alpha, \qquad p\Gamma = p_\alpha \Gamma^\alpha \tag{9}$$

$$\Gamma^\alpha = (\gamma^\alpha_{(1)} + \gamma^\alpha_{(2)})/2 \tag{10}$$

and the x^α are the photon coordinates, $\hbar = 1$, $c = 1$.

Equation (8), (10) is simply Equation (67) of [1 (a), p. 38] in another representation of the wave function. The fundamental properties of the matrices for the unit spin and zero spin particles have been given by Petiau in [3].

From (8), we have after multiplication of both sides by

$$p(\gamma_{(1)} - \gamma_{(2)}),$$

$$p\gamma_{(1)}\Psi = p\gamma_{(2)}\Psi \tag{11}$$

if $\kappa \neq 0$. From (8) and (11),

$$\begin{aligned} &\text{(a)} \qquad p\gamma_{(1)}\Psi = \kappa\Psi \\ &\text{(b)} \qquad p\gamma_{(2)}\Psi = \kappa\Psi \end{aligned} \tag{12}$$

So

$$p^2\Psi = \kappa^2\Psi \tag{13}$$

which shows that κ is the photon mass.

In the ϕ-representation, (12a, b) become

$$\begin{aligned} &\text{(a)} \qquad p\gamma\phi = \kappa\phi \\ &\text{(b)} \qquad -p\phi\gamma = \kappa\phi \end{aligned} \tag{12'}$$

and (8), (11)

$$p(\gamma\phi - \phi\gamma)/2 = \kappa\phi \tag{8'}$$

$$p\gamma\phi = -p\phi\gamma. \tag{11'}$$

The adjoint $\bar{\phi}$ satisfies the same equation. In other words, the system (12′) is invariant under the operation K of complex conjugation

$$Kp = -p, \qquad K\phi = \bar{\phi}.$$

It is also invariant for space (P) and time (T) reflexions, with

$$T\phi(t, \bar{x}) = \gamma_5\phi^+(-t, \bar{x})\,\gamma_5.$$

Using (1), we obtain the tensorial forms of the wave equations. The system (12a)≡(12′a) becomes

$$\begin{aligned}
&i\partial_\alpha A^\alpha=\kappa S\\
&i\partial_\lambda S-\partial^\alpha H_{\alpha\lambda}=\kappa A_\lambda\\
&\partial_\lambda A_\mu-\partial_\mu A_\lambda-i\varepsilon_{\lambda\mu\alpha\beta}(\partial^\alpha B^\beta-\partial^\beta B^\alpha)/2=\kappa H_{\lambda\mu}\\
&\partial_\lambda P-i\partial^\alpha\, {}^xH_{\alpha\lambda}=\kappa B_\lambda\\
&\quad -\partial_\alpha B^\alpha=\kappa P
\end{aligned}\tag{14}$$

where $\varepsilon_{\lambda\mu\alpha\beta}$ is the completely antisymmetrical tensor whose component

$$\varepsilon_{0123}=-1,\ (\varepsilon^{0123}=+1),$$

and

$${}^xH_{\alpha\lambda}=\varepsilon_{\alpha\lambda\beta\mu}H^{\beta\mu}/2.$$

The tensorial form of (12b)–(12′b) is just (14) but with $(-i)$ instead of i. Thus, for (8′)

$$\begin{aligned}
&\text{(a)}\qquad \kappa S=0\\
&\text{(b)}\qquad \left.\begin{aligned}&\partial_\alpha H^{\alpha\lambda}+\kappa A^\lambda=0\\ &\partial_\lambda A_\mu-\partial_\mu A_\lambda=\kappa H_{\lambda\mu}\end{aligned}\right\}\\
&\text{(c)}\qquad \left.\begin{aligned}&\partial_\lambda P=\kappa B_\lambda\\ &-\partial_\alpha B^\alpha=\kappa P\end{aligned}\right\}
\end{aligned}\tag{15}$$

it means three separated systems in which one could put three distinct masses κ_0, κ, κ' instead of κ.

This could also be done in (8′) through a small modification of its right-hand side.

Putting then $\kappa_0=0$

$$\phi_i^{(0)k}=\delta_i^k S/4 \quad \text{with} \quad S=\text{constant}\tag{16}$$

is a solution of the so modified Equation (8′), (11′); it defines the 'state of annihilation' which plays an important role in de Broglie's theory. This state (16) is also a solution of a five dimensional wave equation which generalizes (8′). [1 (c), p. 35 and p. 192].

The vectorial or Maxwellian states, pure unit spin states, are defined by

$$4\phi^{(m)}=A\gamma+H_{\alpha\beta}\sigma^{\alpha\beta}/2\tag{17}$$

or (7). From (15b) it is clear that A_α and $\kappa H_{\alpha\beta}$ are e.m. potentials and fields up to a same factor which for dimensional reasons is equal to $1/\sqrt{\kappa}$ times a numerical constant. More about that at Section 4.

The states defined by

$$4\phi' = B_\alpha \gamma_5 \gamma^\alpha + P i \gamma_5 \tag{18}$$

are pseudoscalar.

Let us note that $p(\gamma\phi + \phi\gamma)/2 = \kappa\phi$ instead of (8′) would give the wave equations for a scalar and pseudo-vectorial particle.

Remarks

The wave mechanics proposed by J. M. Whittaker for the electron has been neglected [4]. His wave equations are simply (14) where p_α should be replaced by

$$p_\alpha + e a_\alpha$$

to take account of an external electromagnetic field. In fact, the J. M. Whittaker particle would be scalar, pseudo-scalar, vectorial and pseudo-vectorial.

Proca proposed much simpler tensorial equations for the electron [5]; as is known, the Proca particle is the vectorial one. For a complete bibliography of Proca's works on that point, see [6, References].

3. Conserved currents

The matrix formalism is well suited to find conservation laws, whatever the spin. Let us give three examples:

(a) From (12) one gets, as for the case of spin $\frac{1}{2}$, the two conserved vector currents

$$\begin{aligned} C^\alpha_{(1)} &= \operatorname{tr} \bar{\Psi} \gamma^\alpha_{(1)} \Psi = -\operatorname{tr} \bar{\phi} \gamma^\alpha \phi \\ C^\alpha_{(2)} &= \operatorname{tr} \bar{\Psi} \gamma^\alpha_{(2)} \Psi = \operatorname{tr} \bar{\phi} \phi \gamma^\alpha \end{aligned} \tag{19}$$

$(\bar{\Psi} = \gamma_0 \Psi^+ \gamma_0)$, or

$$C^\alpha = \operatorname{tr} \bar{\Psi} \Gamma^\alpha \Psi = \operatorname{tr}(-\bar{\phi}\gamma^\alpha\phi + \bar{\phi}\phi\gamma^\alpha)/2 \tag{20}$$

$$J^\alpha = \operatorname{tr} \bar{\Psi} (\gamma^\alpha_{(1)} - \gamma^\alpha_{(2)}) \Psi/2 = -\operatorname{tr} \bar{\phi} (\gamma^\alpha\phi + \phi\gamma^\alpha)/2 \tag{21}$$

In tensor notation,

$$4C_\alpha = iH^*_{\alpha\lambda}A^\lambda + iP^*B_\alpha + \text{c.c.} \tag{20'}$$

$$4J_\alpha = -S^*A_\alpha + {}^xH^*_{\alpha\lambda}B^\lambda + \text{c.c.} \tag{21'}$$

The vector C_α is the vector of photon density and flux. It would be identically zero for real fields. The vector J_α will serve in Section 4.

(b) The first energy-momentum densities which appear in de Broglie's theory are non-Maxwellian, symmetrical or not [1 (a), 1936, (b), 1940, p. 186, (c), p. 16]. Costa de Beauregard has called the attention to one of these non-symmetrical [7a], and used it to interpret properties of evanescent waves because 'the energy flux of the spinning photons is not collinear with their momentum inside Fresnel's evanescent wave' [7 (b)]. Let us remark that the theory of graviton-photon interactions could bring new information on which energy-momentum to choose.

(c) The total angular momentum operator is the sum

$$J^{rs} = L^{rs} + S^{rs} \qquad (r, s = 1, 2, 3)$$

of the orbital momentum

$$L^{rs} = (x^r p^s - x^s p^r)\, I \times I$$

and the spin

$$S^{rs} = \tfrac{1}{2}\sigma^{rs} \times I - \tfrac{1}{2} I \times \sigma^{rs}$$

in the ϕ representation with the notation

$$A \times B\phi = A\phi B.$$

The corresponding conservation laws are

$$\partial_\alpha J^\alpha_{rs} = 0,$$

where

$$J^\alpha_{rs} = -\operatorname{tr} \bar{\phi}\Gamma^\alpha J_{rs}\phi$$

$$\Gamma^\alpha = (\gamma^\alpha \times I - I \times \gamma^\alpha)/2.$$

By this method we obtain quite simply the correct values for the angular momenta carried by plane waves [1, b) Tome I], and by *lm*-polar waves defined as solutions of

$$J_{12}\phi^{(lm)} = m\phi^{(lm)}$$

$$(J^2_{12} + J^2_{23} + J^2_{31})\, \phi^{(lm)} = l(l+1)\, \phi^{(lm)}.$$

For this last case

$$J_{12}^{\alpha}=mC^{\alpha}$$

as required.

Let us note that we have taken here, for the spin, components of

$$S^{\alpha\lambda\mu}=-\bar{\phi}\Gamma^{\alpha}S^{\lambda\mu}\phi.$$

The completely antisymmetrical spin tensor used in [8, p. 118] and [1, c) Ch VI] leads to the same results, but in the particular gauge where $A_0=0$.

4. Electromagnetic interactions

The theory of the electromagnetic interactions can be based on the elementary interaction hamiltonian between a charged particle and one photon. [1, b) Tome II, p. 70 and c) Ch XI]. We shall indicate an equivalent method, in the quantized field theory. From now on, ϕ is a quantized field, and so are the spinors ψ, $\bar{\psi}$ in the electron current

$$j_{\alpha}=\bar{\psi}\gamma_{\alpha}\psi; \tag{22}$$

the star in ϕ^*, ψ^* means now hermitian conjugate.

The S, A_{α}, ..., (S^*, A_{α}^*, ...), are operators of destruction (creation) of annihilated, vectorial, ...photons. Thus, for examples, S^*A produces a transition of a photon from a vectorial to an annihilated state.

To get the Lagrangian density for the electromagnetic interactions in accord with de Broglie's views we have to treat the charged particles – the electrons for example – and the photon on the same level, as far as possible. Taking into account the usual Lagrangian density, we see that we have to take here as the Lagrangian density

$$\varepsilon j^{\alpha}J_{\alpha}; \tag{23}$$

the interaction terms for the two constituents of the photon are of opposite signs, as if they were particles of opposite charges.

The next step is to compare (23) with

$$ej^{\alpha}A_{\alpha}^{(r)} \tag{24}$$

where $A_{(r)}^{\alpha}$ is the real Maxwellian potential. From (21′)

$$-\varepsilon\sqrt{\kappa}(S^*A_{\alpha}+A_{\alpha}^*S)/2\sqrt{\kappa}=eA_{\alpha}^{(r)}. \tag{25}$$

Further, any physical state certainly contains a very large number n_0, let us say an infinite number of annihilated photons per unit volume. Thus, S reproduces the same state multiplied by a very big number which we may put equal to $\sqrt{n_0}$, the phase factor being included in A^*. Then

$$-\varepsilon\sqrt{\kappa n_0}=e. \tag{26}$$

This compels us to regard ε as an infinitely small quantity which cancels the probability of transformation of a photon from a vectorial to a pseudoscalar state and vice-versa. Then, the interaction density (23) gives no possibility of creation of a pseudoscalar photon, and we shall neglect those states in the following.

Let us stress that (25) is obtained from (21) where the quantized ϕ is, in conformity with de Broglie's point of view, an operator of annihilation only.

5. The Photon Rest Mass

The convenience of a massive electrodynamics is generally recognized [9, 6–5, 15–2, 15–4], but for de Broglie a non-zero value of the photon rest mass is essential.

Up to now the best established quantity is an upper limit: $\kappa \leqslant 3\times 10^{-15}$ eV, [6]). As is well known this smallness reduces considerably the real e.m. interactions with a longitudinal photon compared to those with a transverse one. Recently, de Broglie and Vigier [10] have used a massive photon to interpret Imbert's experiments [11] about shifts of position of internally reflected light beams but interpretations with massless photons have been given afterwards [12], [13].

Cosmological events could be more appropriate. Following Vigier [14] collisions between longitudinal and transverse photons might explain observed anomalous red-shifts. On another hand, Deser [15] has noted that through gravitational experiments, longitudinal photons could be detected as well as transverse ones. Other types of interactions with the same behaviour could be imagined. But no definite conclusion can be drawn about the photon mass.

Université de Bruxelles

BIBLIOGRAPHY

[1] de Broglie, L.,
(a) *Act. Sc. et ind.*, Hermann et Cie. Paris, 1934, No. 181 and 1936, No. 411.
(b) *La Mécanique ondulatoire du photon. Une nouvelle théorie de la lumière* Tome I (1940); Tome II (1942).
(c) *Mécanique ondulatoire du photon et théorie quantique des champs*, Gauthier-Villars, Paris, deuxième éd., 1957.
(d) *Théorie générale des particules à spin (Méthode de fusion)*, Gauthier-Villars, Paris, 1943.

[2] Tonnelat, M.-A., Thesis, University of Paris, 1942.

[3] Petiau, G., Thesis, University of Paris, 1936; and *Acad. Roy. Belg. Cl. Sci., Mém. in 8°*, **16** (1936), fasc. 2.

[4] Whittaker, J. M., *Proc. Roy. Soc.* **A121** (1928), 543.

[5] Proca, A., *J. Phys. Rad. Srie VII* **6** (1936), 347, and **8** (1937), 23.

[6] Goldhaber, A. S. and Nieto, M. M., *Rev. Mod. Phys.* **43** (1971), 277.

[7] Costa de Beauregard, *Foundations of Physics* **2** (1972), 125; and Thesis, University of Paris, *Journal de Mathématiques* **22** (1943), 85–176.

[8] Géhéniau, J., *Acad. Roy. Belg., Cl. Sci., Mém. in 8°* **18** (1938).

[9] Jauch, J. M. and Rohrlich, F., *The Theory of Photons and Electrons*, Addison-Wesley Pub.. Co., 1955.

[10] de Broglie, L. and Vigier, J. P., *Phys. Letters* **28** (1972), 1001.

[11] Mazer, A., Imbert, C., and Huard, S., *C. R. Acad. Sci. Paris* **B173** (1971), 592.

[12] Ashby, N. and Miller Jr., S. C., *Phys. Rev.* **D7** (1973), 2383.

[13] Boulware, D. G., *Phys. Rev.* **D7** (1973), 2375.

[14] Pecker, J. C., Roberts, A., and Vigier, J. P., *C.R. Acad. Sci. Paris* **B274** (1972), 765; and *Nature* **241** (1973), 338.

[15] Deser, S., *Ann. Inst. H. Poincaré* **16** (1972), 79.

MARIE ANTOINETTE TONNELAT

FROM THE PHOTON TO THE GRAVITON AND TO A GENERAL THEORY OF CORPUSCULAR WAVES*

1. INTRODUCTION

Dirac's theory of the electron had introduced, as if by accident, the notion of spin into the very centre of quantum mechanics. In suggesting linear relativistic wave equations, P. A. M. Dirac [5] had been able to give an account of the anomalous Zeeman effect and of the characteristics of fine structure. The notion of spin thus appeared to determine even the type of equations which described the behaviour of a given particle. It played a fundamental part in a process of classification of particles.

Starting from Dirac's equations which were valid for the electron, that is for a fermion of rest mass M_0 and spin $\frac{1}{2}$ (in units of $h/2\pi$), we can imagine a process known as 'fusion', associating an even or odd number of such Dirac particles. In the first case, the association (fusion) of $2n$ ($n=1, 2, 3$) particles will lead to the wave equations of a boson of maximum integer spin n; in the second case, the fusion of $2n+1$ Dirac particles will give the description of a fermion with maximum half-integer spin $n+\frac{1}{2}$.

The simplest example clearly consists of postulating the fusion of two Dirac particles. Thus Louis de Broglie obtained in 1934 the equations of a particle with a spin one: the equations for a photon having a mass, that is Maxwell equations completed by terms in μ_0 [9]. These relationships, identical with those to be obtained later by A. Proca using another procedure, can also describe a (then new) spin 1 particle: the meson. (Cf. Géhéniau [9]).

The procedure introduced by Louis de Broglie could of course be generalised. The next step (Tonnelat [15], Petiau [12]) consists of the fusion of 2 photons, that is of 4 Dirac particles, thus obtaining the wave equations for a particle of spin 2: the graviton.

At that time (1939) a general and systematic study of particles with spin appeared to be the key to a fruitful knowledge of matter. In 1935, Petiau had given the characteristic relationships between the operators which described the behaviour of certain types of particle: the Petiau-Duffin-

M. Flato et al. (eds.), Quantum Mechanics, Determinism, Causality, and Particles, 227–235. All Rights Reserved.

Kemmer relationships [11]; in 1938 Pauli and Fierz suggested, within the framework of field theory, equations for particles of any spin whatsoever. Due to the war, this work remained unknown in France, and in addition the syntheses based on the fusion process suggested by Louis de Broglie were rather different in nature [3].

It was not in fact just a wish of generalization that prompted Louis de Broglie to associate two Dirac particles to form a 'heavy photon'. The equations thus obtained have a particular solution known as the 'annihilation solution' which seemed to Louis de Broglie to correspond to a possible description of the photoelectric effect: two material particles (electrons) appeared to be able to vanish (annihilation) to produce an electromagnetic radiation. Lastly, the longitudinal waves associated with the propagation of a photon with nonzero mass played an essential part in the mechanism of Coulomb interactions.

In a similar manner, the physical interpretation permitted by the method of fusion applied to particles of any spin must play a preponderant part in a theory of corpuscular waves. In the first place, it was to show itself in the simplest of these: the theory of the graviton.

2. The Graviton, a Particle with a Spin 2

At the time when Dirac was constructing a relativistic theory of the electron, and when Louis de Broglie was suggesting a relativistic theory of the photon, gravitation appeared to be separated by a very wide gulf from other physical phenomena. Certainly, as was stressed by Louis de Broglie, it could appear paradoxical that the fastest particle, the photon, had not been described by a relativistic treatment until 1936. On the other hand, gravitation appeared to raise quite different problems. In fact, since 1915, Einstein's theory of gravitation, known as 'general relativity' had been accepted by all physicists. According to this theory, gravitation phenomena are a consequence of the non-Euclidean structure of space. In this curved universe created by the masses, the test particles move along geodesics which coincide, to a first approximation, with the orbits of the celestial bodies.

The geometric interpretation of gravitation is one of the simplest and most elegant theories conceivable. It must be admitted, however, that it separates gravitation sharply from other physical phenomena, and in particular from electromagnetism, with which it had hitherto always been to some extent associated.

There is of course no incompatibility between general relativity and corpuscular wave equations. If true space is a Riemann space, we can easily write out wave equations valid in such a space. Thus Schrödinger [13], Pauli [10] generalized Dirac's equations to a non-Euclidean space. In the same way, the equations of the photon can be extended to a curved space [14].

Nevertheless, these generalizations show only that there is a compatibility between two explanatory themes, the roles of which remain fundamentally different: in quantum mechanics, curved space remains a permissible framework; according to general relativity it becomes an effective cause. In the first case, the corpuscular wave equations are valid in the Riemann space; in the second case, this space becomes responsible for the nature of the orbits. It therefore seems that a form of option becomes necessary: either to geometrize all the forms of energy (as is the aim of the theories known as 'unitary'), or, on the contrary, to interpret gravitational energy by a specific quantum phenomenon. Eddington had already written in 1923 that "the mechanical properties of matter are commonly represented by the curvature of space-time" [6]; but, if we prefer, we can resolve this into particles and represent the mechanical properties by wave functions. We cannot consider it both ways at the same time" [6].

Since the mainly philosophical speculations of Boscovitch and of Le Sage (extra-terrestrial corpuscles), no successful effort had been made to describe gravitational phenomena by a corpuscular process. If we wish to make use of the fusion process of Louis de Broglie, we easily find that the association of two Dirac particles introduces quantities dependent on a quadrivector (spin 1) linked with the quadripotential φ_μ; the fusion of four Dirac particles makes all the field quantities relative to the maximum spin 2 dependent on a symmetrical tensor of the second rank $\phi_{(\mu\nu)}$. This tensor is clearly the analogue of the non-Euclidean part of the metric tensor of general relativity. Its behaviour is described by linear equations comparable to the linear approximation of general relativity.

3. Fusion of 2 photons and linearized gravitational equations

More precisely, two Dirac particles described by the two wave functions with 4 components $\psi_i^{(1)}$ and $\psi_m^{(2)}$ $(i, m=1, 2, 3, 4)$ satisfy the equations

$$(\gamma^\mu \partial_\mu - k_0)\,\psi_i = 0 \qquad (\gamma^\mu \partial_\mu - k_0)\,\psi_m = 0 \tag{1}$$

with

$$\left(\partial_\mu = \frac{\partial}{\partial x^\mu}, \qquad k_0 = \frac{2\pi}{h}\,\mu_0 c\right)$$

Two other corpuscles $\varphi_k^{(1)}$ and $\varphi_k^{(2)}$ satisfy the corresponding equations

$$\partial_\mu \varphi_l \gamma^\mu + k_0 \varphi_l = 0 \qquad \partial_\mu \varphi_k \gamma^\mu + k_0 \varphi_k = 0 \tag{2}$$

The wave function

$$\Psi_{iklm} = \psi_i^{(1)} \varphi_l^{(1)} \psi_m^{(2)} \varphi_k^{(2)} \tag{3}$$

has $4^4 = 256$ components and satisfies the linear equations

$$(P^\mu \partial_\mu - k_0)\,\Psi = 0 \tag{4}$$

$$Q^\mu \partial_\mu \Psi = 0, \qquad R^\mu \partial_\mu \Psi = 0, \qquad S^\mu \partial_\mu \Psi = 0. \tag{5}$$

The operators P, Q, R, S are formed by linear combinations of four Dirac operators acting respectively on one of the four indices of the wave function. It can be shown that (4), in conjunction with any one of equations (5) justifies the validity of the whole system.

The function Ψ_{iklm} can be regarded as a matrix with 16 lines and 16 columns that can be expressed in terms of a basic system of matrices of this type. Thus, generalizing the expression of the principle of fusion, we will put

$$\Psi_{iklm} = [(\gamma^A)_{il}\,(\gamma^B)^+_{mk} \pm (\gamma^A)^+_{mk}\,(\gamma^B)_{il}]\;\phi_{AB} \tag{6}$$

where the sign $^+$ indicates the conjugate matrix, and $A, B = 1, \ldots, 16$.

The physical field ϕ_{AB}, quantities which are clearly complex, are thus divided into 4 groups: one corresponds to pure spin $j=2$, three to the state where spin $j=1$, and two to spin $j=0$ (Tonnelat [15]). Let us write three of these groups, the two others being identical to Maxwell's equations completed by mass terms

$$\begin{cases} \partial_\mu \phi_{(\nu\rho)} - \partial_\nu \phi_{(\mu\rho)} = k_0 \phi_{[\mu\nu]\rho} \\ \partial^\rho \phi_{[\rho\mu]\nu} = -k_0 \phi_{(\mu\nu)} \\ \partial_\mu \phi_{[\rho\sigma]\nu} - \partial_\nu \phi_{[\rho\sigma]\mu} = k_0 \phi_{([\mu\nu][\rho\sigma])} \end{cases} \tag{S$_2$}$$

$$\begin{cases} \partial_\mu \psi_{[\nu\rho]} - \partial_\nu \psi_{[\mu\rho]} = k_0 \psi_{[\mu\nu]\rho} \\ \partial^\rho \psi_{[\rho\mu]\nu} = k_0 \psi_{[\mu\nu]} \\ \partial_\mu \psi_{[\rho\sigma]\nu} - \partial_\nu \psi_{[\rho\sigma]\mu} = k_0 \psi_{[[\mu\nu][\rho\sigma]]} \end{cases} \tag{S$_1$}$$

$$\begin{cases} \partial_\mu \phi^{(0)} = k_0 \phi^{(0)}_\mu & \phi^{(0)} = \sum_\rho \phi_{(\rho\rho)} \\ \partial^\rho \phi^{(0)}_\rho = -k_0 \phi^{(0)} \quad \text{with} & k_0 \phi^{(0)}_\mu = k_0 \sum_\rho \phi_{[\mu\rho]\rho} = \partial_\mu \phi^{(0)} \\ \partial_\mu \phi^{(0)}_\nu = k_0 \phi^{(0)}_{\mu\nu} \end{cases} \tag{S$_0$}$$

The group (S_0) is obtained from (S_2) by contraction. Then by putting

$$\phi^{(2)}_{\mu\nu} = \phi_{(\mu\nu)} - \tfrac{1}{3}\delta_{\mu\nu}\phi^0 - \tfrac{1}{3}\phi^{(0)}_{\mu\nu} \tag{7}$$

we can obtain equations for a particle of total pure spin $2(\sum_\rho \phi^{(2)}_{(\rho\rho)} \equiv 0)$, without mixture of spin $j=0$. The equations satisfied by the quantities $\phi^{(0)}$ are identical with (S_2) (Cf. van Isacker [2]).

We can also show [2] that the equations (S_1) reduce to Maxwell's equations completed by mass terms.

It is easily established (M. A. Tonnelat [15], [16]) that the equations (S_2), valid in vacuum, are identical with the linear approximation of Einstein's equations, these latter now being written with a cosmological term. In particular, the tensor $\phi_{([\mu\nu][\rho\sigma])}$ has the symmetry features of the Riemann-Christoffel tensor. It is formed with the $\phi_{[\mu\nu]\rho}$ and the $\phi_{(\mu\nu)}$ as the Riemann tensor is formed from the Christoffel symbols and the components of the metric tensor. It can even be shown that the three groups (S_2), (S_1) and (S_0) constitute the linear approximation of the equations of the asymmetric Einstein-Schrödinger theory, this latter also including the cosmological term mentioned by Schrödinger [16].

4. Coulomb and Newton Interactions

In constructing his theory of the photon, Louis de Broglie had shown the existence of plane waves, of which certain spin components in a given direction, $m=1$, correspond to right-hand polarized and left-hand polarized components and others, $m=0$, to a purely longitudinal polarization. According to Louis de Broglie, these latter components ensure the Coulomb interactions between two charged particles. Indeed, the interaction term

$$K_c = L_e j^\alpha A_\alpha \qquad L_c = \text{arbitrary constant} \tag{8}$$

introduces the density-current quadrivector of Dirac's theory and the electromagnetic potential A_α created by the photon. Postulating that the interaction between two charged particles takes place as a result of the longitudinal waves alone, we can calculate the elements of the matrix H_{if} of an interaction Hamiltonian causing the transition of a system of two charged particles from the initial state (i) to the final state (f) by emission and absorption of longitudinal photons.

The interactions between charged particles by photon exchange thus constitute, as is stated by Louis de Broglie, a "double transition, resulting overall in the transfer of an electron from one state of annihilation to another state of annihilation". The intermediate state of the system is a virtual state and it is globally (that is from initial state to final state) that we can have conservation of energy and momentum.

This principle, assumed by Louis de Broglie in the case of the photon, may be extended to the spin 2 particle. Two material particles exchange the 'longitudinal' gravitons corresponding to $m=0$ (longitudinal waves). We can then determine the symmetric potential $\phi_{(\alpha\beta)}$ corresponding to the only longitudinal actions. The interaction Hamiltonian

$$K_g = L_g T^{\alpha\beta} \phi_{(\alpha\beta)} \qquad (L_g = \text{arbitrary constant}) \tag{9}$$

($T^{\alpha\beta}$ is the energy momentum of the material particles) is easily deduced from this. By making the traditional assumptions, we find that the elements of the matrix so calculated correspond to a potential

$$U = -G \frac{m_1 m_2}{|\mathbf{r}_{12}|} e^{-k_0 |\mathbf{r}_{12}|} \tag{10}$$

The potential U differs from the Newtonian potential by a Gaussian function. This latter is clearly nearly equal to unity if the mass of the graviton is very small.

We thus return to a sort of symmetry between gravitation and electromagnetism. The strange similarity between Coulomb and Newton forces, together with their sign difference, which had raised so much difficulty in post-Newtonian physics, may thereby receive the beginnings of an explanation. Unification, however, is no longer placed under the heading of geometry, but under that of the similar behaviour of particles carrying charges or masses. By exchange of photons or of gravitons, they carry out transitions, the macroscopic summation of which is described either by Coulomb's law if there is an exchange of photons, or by Newton's law if there is an exchange of gravitons. It is for this reason that we can insist on this aspect of the question by giving this theory the title "A new form of unitary theory. Study of the spin 2 particle" [15].

Many questions can of course be asked: the transition from complex microscopic quantities to real macroscopic ones, the very meaning of 'state of annihilation', the transition to a non-linear condition, the study of fields not in vacuum, but in matter; all is not yet made clear. However, it seems clear that the method of fusion, by introduction of non-zero rest masses, cannot regard the abolition of these masses as a simple special case or as a limiting value (de Broglie and Tonnelat [4]). In this sense, the possibility of according a physical significance to the potentials was used as a criticism of this theory at a time when gauge invariance was a sacred dogma of electromagnetism.

5. Particles of arbitrary spin

By generalizing the theories of the photon and the graviton, we could easily formulate wave equations for bosons of arbitrary integer spin. In the same way, fusion of an odd number of Dirac particles led to wave equations for fermions. These extensions (Cf. Petiau [11, 14]) have been analyzed and frequently completed by Louis de Broglie in his *General Theory of Spin Particles* [2]. This synthesis remains faithful to the physical principles which had been the basis of the 'new theory of light': the association of a physical field with every particle, and the statement of the correspondence between the field quantities and the corpuscular behaviour of the particle which remains fundamental. Study of plane waves and of the energy-momentum tensor(s) is based on this same principle. We can of course ask ourselves whether the general study of particles with spin supposes that effective complex structure which the theory seems to postulate. We can also give thought to the fact that we are dealing here with a first linear approximation, which could lead to a non-linear extension. Once again, general relativity would be both an obstacle and a model. Questions of this nature were asked by Louis de Broglie in the last chapters of his book on the general theory of particles with spin [2]. Such also are the problems which he developed in his *Wave Mechanics of the Photon and Quantum Theory of Fields* [3].

"It appears certain to us", wrote Louis de Broglie, "that this theory [referring to the wave mechanics of the photon and its extension to particles with spin] has the great merit of demonstrating clearly the true physical meaning of the fairly abstract formulation of the quantum theory of fields and of giving a clear indication on questions which remain fairly obscure". This merit does not of course enable Louis de Broglie to solve all mysteries, but, without avoiding them, it provides a means of suggesting a clear statement of them, which is too frequently forgotten, and of genuine search for an interpretation. Such was the aim of the first relativistic theory of light, hopes linked to a unitary theory of corpuscular type and subsequent generalizations. Whilst the questions raised are now stated in slightly different terms, they are very far from being answered.

Université de la Sorbonne, Faculté Pierre et Marie Curie

NOTE

* Translated from the French by N. Corcoran.

BIBLIOGRAPHY

[1] de Broglie, Louis, *L'électron magnétique*, Théorie de Dirac, Hermann, Paris 1934.
[2] de Broglie, Louis, *Théorie générale des particules à spin – Méthode de Fusion*, Gauthier-Villars, 1954.
[3] de Broglie, Louis, *Mécanique ondulatoire du photon et théorie quantique des champs*, Gauthiers-Villars, 1956.
[4] de Broglie, Louis, and Tonnelat, M. A., 'Sur les possibilités d'une structure complexe pour la particule à spin 2', *C.R. Acad. Sci.* **230** (1950), 1329.
[5] Dirac, P. A. M., 'The Quantum Theory of the Electron', *Proc. Roy. Soc.* **A117**, 1928, p. 610.
[6] Eddington, A. S., 'Electrons and Photons'.

[7] Fierz, M., 'Ueber die relativistische Theorie kraftfreier Teilchen mit beliebigem Spin', *Hel. Phys. Acta* **12** (1939), 3.
[8] Fierz, M. and Pauli, W., 'On Relativistic Wave Equation for Particles of Arbitrary Spin in an Electromagnetic Field', *Proc. Roy. Soc.* **173A** (1939), 211.
[9] Géhéniau, J., 'Mécanique ondulatoire de l'électron et du photon', *Bull. Acad. Roy. Belg.* (1937).
[10] Pauli, W., 'Ueber die Formulierung der Naturgesetze', *Ann. der Phys.* **18** (1933), 337.
[11] Petiaŭ, G., 'Contributions à l'étude des équations d'onde corpusculaires', *Bull. Acad. Roy. Belg.* **76** (1936), 118.
[12] Petiau, G., 'Sur les équations d'ondes des corpuscules de spin quelconque', *J. Phys. Rad.* **7** (1946), 124.
[13] Schrödinger, E., 'Diracches Elektron im Schwerfeld', *Sitz Preuss Akad. d. Wiss.* **3** (1932), 205.
[14] Tonnelat, M. A., 'Sur la théorie du photon dans un espace de Riemann', *Ann. der Phys.* **15** (1941), 144.
[15] Tonnelat, M. A., 'Une nouvelle forme de théorie unitaire – Etude de la particule de spin 2', *Ann. der Phys.* **17** (1942), 151–208.
'Etude des interactions entre la matière et la particule de spin 2', *Ann. de Phys.* **19** (1944), 396–445.
[16] Tonnelat, M. A., *Les théories unitaires de l'électromagnétisme et de la gravitation*, Gauthier-Villars, 1965.

JEAN-PIERRE VIGIER

POSSIBLE IMPLICATIONS OF DE BROGLIE'S WAVE-MECHANICAL THEORY OF PHOTON BEHAVIOUR

Analyzing in a recent publication [1] the historical origin of Wave Mechanics, Professor de Broglie has recalled that its origin rests on the almost forgotten idea that all particles are endowed with an intrinsic oscillation: so that one can compare them to clocks which remain in phase with the ψ waves utilized in Quantum Theory. This postulate of phase correlation is of relativistic nature. It yields an immediate demonstration of the identity of the principles of Fermat and Maupertuis.

When applied to the theory of light this idea provides a simple explanation of an experimental property which has never been explained in the usual theory. If one considers the action of hertzian waves (of frequency ν) on an oscillating circuit the absorbed energy is evidently split into isolated quantas which yield a discrete set of isolated impulsions. A regular oscillation can then only be permanently maintained if these impulsions are correlated with the frequency of the circuit which corresponds to the frequency of the incident wave ... so that individual photons behave as if they carried individual frequencies equal to ν.

If we accept this idea we see also that the analogy between the theory of light and particle theory is only complete if we endow individual photons with a non-zero rest mass $\mu \leqslant 10^{-48}$ gr.

The existence of such a mass has been shown [2] to be compatible with recent experimental data and with currently known consequences of Quantum Electrodynamics.

The aim of the present paper, written in homage to Louis de Broglie, is to follow up some possible theoretical and experimental consequences of this latter assumption ($\mu \neq 0$) and to see whether de Broglie's original idea can help us to interpret puzzling new experimental data (such as possible 'anomalous' redshifts in astronomical observations [3] or beams of parallel high energy γ's observed in cosmic ray plates [4] and to attack well-known unresolved theoretical difficulties in the present theory of light.

M. Flato et al. (eds.), Quantum Mechanics, Determinism, Causality, and Particles, 237–249. *All Rights Reserved. Copyright © 1976 by D. Reidel Publishing Company, Dordrecht-Holland.*

A. The first theoretical point is that $\mu \neq 0$ determines the numerical value of $\alpha = e^2/\hbar c$.

To show this following Adler and Bardeen [5] we can now work out free electron theory in a constant external electromagnetic field with a non-zero photon mass which can be taken equal to its true basic mass. This means that we neglect all photon self-energy graphs and work in the transverse Landau gauge.

As shown by Adler and Bardeen [5] if we denote by e_0 the basic electron charge and m the electron mass, the asymptotic electron propagator for renormalized wave functions takes the form:

$$\tilde{S}'_F(p)^{-1}_{\text{Landau gauge}} \underset{p \to b}{\sim} c\left[\gamma \cdot p + am\left(\frac{m^2}{-p^2}\right)^{\varepsilon}\right] \tag{1}$$

in the infinite momentum frame (I.M.F.) where $\rho^2 \gg m^2$, with $\varepsilon = (1/2)\,\alpha(e_0)$ and $\alpha_0 = e_0^2/4\pi$. The corresponding internal photon propagator can be written

$$e^2 \tilde{D}'_F(q)_{\mu p} \underset{q^2 \gg p^2}{\sim} -g_{pr}\frac{e_0^2}{q} + \text{gauge terms} \tag{2}$$

and we can show that the physical value of α can be deduced from its asymptotic value α_0 from the relation $\alpha = \alpha_0$, α_0 being the singular eigenvalue of Gell-Mann and Low's equation $\rho(\alpha_0) = 0$.

We deduce immediately from this the asymptotic form of the corresponding Green function for an isolated electron moving in its own electromagnetic field in a constant external vector potential $B\mu$. Indeed introducing, in Feynman's notation [6] the function of $x = \gamma^\mu x_\mu$ and $B = \gamma^\mu B_\mu$ i.e. $\phi_{e^2}(B, x)$ which represents the amplitude that an electron will arrive in x_μ in its own electromagnetic field (without leaving any photon in the external field) one deduces from Feynman's relation ($\nabla = \gamma^\mu \cdot \partial/\partial x_\mu$):

$$(i\nabla - B)\,\phi_{e^2}(B, x) = ie^2\gamma_\mu \int \delta_+(S^2_{x_1})\left[\frac{\delta\phi_{e^2}(B, x)}{\delta B_\mu(1)}\right] \mathrm{d}\tau_1$$

the exact solution

$$\langle x_{i+1} \mid x_i \rangle^B_{\delta t} = \langle x_{i+1} \mid x_i \rangle^0_{\delta t} \cdot \exp[-i(\delta t \cdot B_4 - (x_{i+1} - x_i)\, B_i)]$$

which represents the transition form x_i to x_{i+1} if we split its path in infinitesimal intervals of δt. This is evidently multiplied by $e^2/4\pi$ (in the system where $\hbar = c = 1$) since the factor appears once (see relation (2)) each time the moving electron emits and absorbs a photon between the two points.

This yields the following theorem:

THEOREM I. If we suppress all photon self-energy graphs in Quantum Electrodynamics (henceforward included in a non-zero photon mass) the propagator of a charged Dirac particle ($m \neq 0$) converges asymptotically (in the I.M.F. where $p^2 \gg m^2$) towards the propagator of a zero-mass fermion. The associated Green function in momentum space then satisfies a wave equation which is invariant under the conformal group $C(M^4) = SO(4, 2) \sim SU(2, 2)$. The constant α then appears as a coefficient of this asymptotic function.

We shall now discuss this asymptotic form in more detail through the utilization of a five-dimensional formalism. More precisely, we are going to show that the preceding I.M.F. motions (and associated α definition) can be mapped on the I.M.F. motion of a Dirac uncharged particle in a particular five-dimensional space. This implies that α now appears as a multiplying factor of the asymptotic Green function in five-dimensional momentum space i.e.

THEOREM II. The asymptotic motion of a Dirac particle treated within the frame of Theorem I (and the corresponding α definition) can be mapped on the motion of a neutral Dirac particle in a five-dimensional fiber space U_5: the fifth space-like dimension corresponding to a closed circle of radius ξ at each point. This satisfies locally the SO (4, 1) invariant wave equation ($\hat{\partial}_i$ representing covariant derivatives):

$$(\gamma^i \cdot \hat{\partial}_i + a)\, \psi = 0 \tag{3}$$

which reduces locally to the SO (5, 2) invariant relation

$$\gamma^i \partial_i \psi = 0 \tag{4}$$

for suitably chosen co-ordinates. The constant α then appears as factor multiplying the Fourier transform, in momentum space, of the elementary solution of Equation (4).

Its demonstration needs no development since its elements have been already discussed in the physical and mathematical literature, in particular by Souriau [7]. Using the latter's notations we define a five-dimensional Riemannian fiber space U_5 where the fifth spacelike dimension behaves like a closed loop (fiber) of radius ξ at each point.

We introduce a principle of relativity which treats the five dimensions of U_5 on an equal basis. The electromagnetic four-vector potential $\mathscr{A}_\mu$ is defined as $\mathscr{A}_\mu = g_{\mu 5}/g_{55}$ where $i=1,\ldots,4$ and we use a local signature $(- - - - +)$.

Then writing $\gamma^i \cdot \gamma^k + \gamma^k \cdot \gamma^i = 2g^{ik}$ ($\hat{\partial}_j$ denoting the covariant derivation of the spinors in the direction j) we see immediately that the generalized Dirac equation in U_5, i.e.

$$(\gamma^i \cdot \hat{\partial}_i + a)\, \psi = 0 \tag{3}$$

(a constant), becomes invariant under the conformal group in five dimensions SO (5, 2) for $p^2 \gg a^2$. So that (3) can be written in the form $\gamma^i \cdot \partial_i \psi \simeq 0$ (4) in the asymptotic I.M.F. limit. This limit corresponds to the asymptotic motion (1) since Souriau has shown that (3) reduces to the Dirac equation for massive charged particles with the correct minimal coupling $eA_\mu j^\mu$. Indeed at each point we can introduce transverse variables in which the hypersurfaces $x^5 = \text{const}$ are orthogonal to the fibers. So we write (the symbol $\sim$ denoting that we are working in these variables)

$$[\tilde{\gamma}_\mu, \tilde{\gamma}_\nu] = 2\tilde{g}_{\mu\nu},\ \tilde{\gamma}^\mu = \tilde{g}^{\mu\nu}\tilde{\gamma}_\nu,\ [\tilde{\gamma}_\mu, \tilde{\gamma}_5] = 0,\ \tilde{\gamma}_5^2 = -\xi^2$$

$$\tilde{\gamma}^\mu = \gamma^\mu,\ \gamma^5 = \tilde{\gamma}^5 - (e/\hbar)\,\tilde{\gamma}^\mu A_\mu,\ \gamma_\mu = \tilde{\gamma}_\mu + (e/\hbar)\,\tilde{\gamma}_5 A_\mu,\ \gamma_5 = \tilde{\gamma}_5\,.$$

When we introduce these in the developed form of (3) i.e.

$$\gamma^i \partial_j \psi + \left\{ \frac{1}{2\sqrt{|g|}}\, \partial_j [\sqrt{|g|} \cdot \gamma^j] - \tfrac{1}{8}\gamma^j [\partial_j \gamma_k - \partial_k \gamma_j]\, \gamma^k \right\} \psi + a\psi = 0$$

we get (expanding ψ in a Fourier series $\psi = \sum_z \psi_z(x_\mu) \exp[iZx^5]$ Z being an integer, and introducing $\Gamma^\mu = \tilde{\gamma}^\mu$, $\Gamma_5 = \tilde{\gamma}_5/\xi = \Gamma_1\Gamma_2\Gamma_3\Gamma_4$ the relation

$$\Gamma^\mu \left[\partial_\mu - \frac{iZ \cdot e}{\hbar} A_\mu \right] \psi_z + \left[a - \frac{iZe}{\hbar} \cdot \sqrt{\frac{2\pi}{\chi}} \cdot \Gamma_5 \right] \psi_z +$$

$$+ \tfrac{1}{8} \sqrt{\frac{\chi}{2\pi}} \cdot F_{\dot{\mu}\dot{\nu}} \cdot \Gamma^{\cdot\mu} \Gamma^{\cdot\nu} \Gamma^5 \psi_z = 0$$

where χ denotes the usual constant in $\tilde{R}_{\mu\nu} - \tfrac{1}{2}\tilde{R} \cdot g_{\mu\nu} = \chi \tilde{T}_{\mu\nu}$ and $A_\mu =$

$=\xi(2\pi/\chi)^{1/2}\mathscr{A}_\mu$. The last term of (4) can be neglected since $\xi=e^{-1}(\chi/2\pi)^{1/2}$ i.e. $\xi=3.782\times10^{-32}$ cm so that, if we write $m^2=(Ze(2\pi/\chi)^{1/2})^2-a^2\hbar^2$ with $m\lambda=Ze(2\pi\hbar)^{1/2}-aK$ and introduce:

$$\varphi=\frac{1-i\Gamma_5}{2}\psi_z+i\lambda\cdot\frac{1+i\Gamma_5}{2}\psi_z$$

relation (4) reduces to

$$\Gamma^\mu\left(\partial_\mu-\frac{iZe}{\hbar}A_\mu\right)\varphi+\frac{im}{\hbar}\varphi=0 \tag{5}$$

which takes the asymptotic form

$$\Gamma^\mu\left(\partial_\mu-\frac{iZe}{\hbar}A_\mu\right)\varphi=0 \tag{6}$$

in the limit $p^2\gg m^2$.

This completes our demonstration: since, if we work in the local coordinate system where $\hat{\partial}_i\to\partial/\partial x^i$ the equivalence of (3) and (6) implies (with the assumptions and restrictions of our first part) that $e^2\,(\hbar=c=1)$ appears as a factor multiplying the Fourier transform in momentum space of the elementary solution of the asymptotic five-dimensional Dirac Equation (4).

We shall now calculate explicitly this coefficient using well-known mathematical results of Hua [8] and Wyler [9]. We can indeed demonstrate.

THEOREM III. The coefficient of the Fourier transform of the elementary solution on $\mathrm{CH}(D^5)$ of relation (4) invariant under $C(M^5)\sim\mathrm{SO}(5,2)$ can be written in the form

$$f=(4\pi)^{-3/2}(2\pi)^{-5/2}\cdot V(Q^5)^{-1/2}\cdot V(D^5)^{1/4} \tag{7}$$

if one denotes by $V(Q^5)^{-1/2}$, $V(S^4)$ and $V(D^5)$ the respective volume of the spaces D^5 (interior of a complex five-dimensional sphere of radius one) Q^5 its Silov boundary and S^4 the surface of a unit sphere in a five-dimensional real Euclidian space.

This can be shown in the following way. At each point $\in U_5$ one introduces a tangent Euclidian space M^5 and the associated spinor fields defined as usual on $M^5 = P(M^5)/SO(4, 1)$; $P(M^5)$ denoting the local Poincaré group. These spinors are eigenfunctions of the associated Laplace-Beltrami operator with eigenvalue O. Since we have in general $\Box_n E_n = (4\pi)^{(n-2)/2} \cdot \delta \cdot I_n$ where I_n denotes the $n \times n$ unit matrix one can show that the elementary solution of equation (4) defined by the relation $(\gamma_i \partial^i)\, S_5 = (4\pi)^{(5-2)/2}\, \delta I_4$ is related to the elementary solution E_5 of $\Box_5 E_5 = (4\pi)^{(5-2)/2} \cdot \delta \cdot I_4$ through the relation $S_5 = (\gamma^i \partial_i)\, E_5 \cdot I_4$. Its Fourier transform $S_5(k)$ defined by the relation

$$S_5(k) = \left(\frac{1}{2\pi}\right)^{5/2} \int S_5(x) \exp(i\langle k \mid x\rangle)\, \mathrm{d}x, \ldots \mathrm{d}x_5$$

also satisfies $(\gamma^i \partial_i)\, S_5(k) = I_4$ so that one obtains the relation

$$S_5^2(k) = E_5(k)\, I_4; \tag{8}$$

which connects the elementary solutions (i.e. the multiplicating coefficients) of the Dirac and Laplace operators on M_5. One can calculate them explicitly by mapping M^5 the spinor fields and relation (4) on the bounded representation D^5 of the homogeneous quotient spaces

$$D^5 = SO(5, 2)/SO(5) \times SO(2).$$

To achieve this result one can utilize an intermediate step i.e. the representation of M^5 on the five-dimensional complex space $T^5 = R^5 + iV^5$ where V^5 is the internal part of the forward oriented light cone, and R^5 the five-dimensional euclidian space. The group $L(T^5)$ of linear transformations on T^5 is isomorphic to $P(M^5)$ on T^5.

One then utilizes the stereographic projection $F: T^5 \to D^5$ (which yields an analytic mapping of T^5 on D^5) introduced by Wyler. As one knows its Silov boundary $Q^5 = S^4 \times S^1$ is just the quotient space $SO(5, 2)/P(M^5) \times D$, D representing the dilatation group. The representation $C(M^5) \sim SO(5, 2)$ on the space of solutions $\{\varphi / \Box_5 \varphi\}$ of the wave equation is equivalent to the representation of SO (5, 2) in the space $H(D^5)$ of homomorphic functions on D^5. The Poisson kernels P_5 correspond to the Green functions of a zeromass scalar field.

We have therefore mapped the spinor fields on M^5 into the unitary representations of $C(M^5)$ constructed on the set of all holomorphic eigenfunctions of the Laplace-Beltrami operator (deduced from the

Casimir operator $g_{ij}p^i p^j$) with eigenvalue zero constructed with the invariant metric on D^5. Their set spans a representative Hilbert space $CH(D^5)$ with square integrable matrix elements.

One obtains finally in this way the relation

$$S_5(z, \xi) = P_5(z, \xi) \tag{9}$$

z and ξ representing the variables on D^5 and Q^5. With (8) this relation (9) connects the elementary solutions of the Dirac and Laplace operators on D^5. Since Hua [8] and Wyler [9, p. 22] have shown that the factor which multiplies S_5 in D^5 can be written $V(Q^5)^{-1}\ V(D^5)^{1/2}$ we see that the factor which multiplies S_5 in D^5 can be written $V(Q^5)^{-1/2}\ V(D^5)^{1/2}$ which shows that the coefficient of the Green function in the momentum space of D^5 takes the value of Theorem III i.e.

$$f = (4\pi)^{-3/2} \cdot (2\pi)^{-5/2} \cdot V(Q^5)^{-1/2} \cdot V(D^5)^{1/4}.$$

The associated value in T^5 being $V(S^4)$. f if one utilizes [9 p. 22] the relation established by Wyler which connects the holomorphic functions on Q^5 and T^5.

The last step in the determination of α is to compare the f value of Theorem III with the value of the coefficient f of the Green function in $D^4 = \mathrm{SO}(4, 2)/\mathrm{SO}(4) \times \mathrm{SO}(2)$ which results from the successive steps

$$M^4 \to T^4 = R^4 + iV^4 \to D^4$$

since according to Theorem I α multiplies the Green function associated to a zero mass particle in four-dimensional momentum space. We thus obtain (substituting 4 to 5 in the demonstration of Theorem III)

$$k = \alpha \cdot (4\pi)^{-1/2} (2\pi)^{-3/2} \cdot V(S^3)^{-1} \cdot V(Q^4)^{-1/2} \cdot V(D^4)^{+1/2}$$

since one passes from five-dimensional motion to four-dimensional motion i.e. from M^5 to M^4 by 'freezing' X^5 the f value of Theorem III can be compared with the value of the coefficient f' of the Green function in the momentum space of $D^4 = \mathrm{SO}(4, 2)/\mathrm{SO}(4) \times \mathrm{SO}(2)$ which results from the successive mappings

$$M^4 \to T^4 = R^4 + iV^4 \to D^4.$$

Now according to Theorem I since α multiplies the Green function in four-dimensional momentum spade we see (by substituting 4 to 5 in the demonstration of Theorem III) that

$$f' = \alpha \cdot (4\pi)^{-1/2} (2\pi)^{-3/2} \cdot V(S^3)^{-1} \cdot V(Q^4)^{-1/2} \cdot V(D^4)^{1/2},$$

where

- $(4\pi)^{-1/2}$ corresponds to the normalization correspondence for holomorphic functions in Q^4 and T^4.
- $(2\pi)^{-3/2}$ results from the four-dimensional Fourier transform.
- $V(S^3)^{-1}$ results from the $Q^4 \to T^4$ transition.
- $V(Q^4)^{-1/2}$ is the P^4 norm factor.
- $V(D^4)^{1/2}$ normalizes to 1 the wave function in D^4.

Writing then $f = f'$ we obtain

$$\alpha = (4\pi)^{-1} (2\pi)^{-1} (2\pi^2) \left(\frac{2^3\pi^2}{3}\right)^{-1} (2\pi^3)^{1/2} \times$$

$$\times \left(\frac{2^3\pi^3}{3}\right)^{-1/2} \left(\frac{\pi^4}{2^6 \cdot 3}\right)^{-1/2} \cdot \left(\frac{\pi^5}{2^4 \cdot 5!}\right)^{1/4}$$

$$= \frac{1}{137{,}037}$$

B. If we apply the principle of phase correlation discussed at the beginning of this paper to neutrinos we must endow both ν_e and ν_μ with a small rest mass m_ν. It is known that we can then construct by fusion non-zero mass γ photons $(J=1)$ and pseudoscalar particles φ $(J=0)$ with $m_\varphi \neq 0$ corresponding to the singlet state. Such particles have been introduced in the literature as quanta associated with the dilatation operator of the conformal group SO (4, 2) $\sim$ SO (2, 2).

If they exist the detection of such particles is only possible (as in the neutrino case) by indirect means: i.e. if we observe reactions where they must be introduced to ensure the conservation of energy momentum and spin.

It is then tempting to associate them with 'anomalous' redshift observations discussed in astrophysics by Arp [3] and various authors, which have also been recently discovered in solar phenomena.

We only mention three:

(a) the 'anomalous' redshift $\delta z = \delta\nu/\nu \sim 10^{-7}$ observed by Roddier [10] one the line $S_n I$ ($\lambda = 4697$ Å) at the edge of the sun;

(b) the anomalous redshift $\delta z \sim 2 \times 10^{-7}$ observed by Goldstein [11] on the wavelength 2292 MHz at the superior conjunction of the satellite

Pioneer-6 behind the sun. This shift analyzed by Merat *et al.* [12] is symmetrical with respect to center of the sun and can be interpreted as an interaction between a φ flux originating from the edge of the sun and the 2292 MHz photons;

(c) the difference $\Delta\gamma$ observed between the axis of the sinusoids corresponding to the Doppler motion of double stars. These axes should superpose in principle. An analysis by Kuhi *et al.* [13] shows it always operates in favour of the hottest component ... and is difficult to interpret with the help of Doppler or gravitational effects.

The interpretation of these 'anomalous' shifts can be attempted within the frame of various theories. The simplest way is to assume the lost energy $\hbar\delta\nu$ If the solar phenomena is carried away by new light particles with strongly interact with the observed γ's. If one utilizes φ to that effect we shall now show it is possible to introduce interactions which satisfy the three main conditions imposed by the present experimental knowledge i.e.

(1) a strong forward scattering peak in the $\gamma-\varphi$ collisions to ensure the point-like character at distant sources. A simple calculation shows that $\langle\delta\varphi\rangle \leqq 10^{-11}$ rad per collision;

(2) a constant fractional energy loss for 10^{10} $H_z < p \leqq 10$ H_z since the observed 'anomalous' shifts are of the same order within these limits;

(3) the absence of consequences which would contradict known results of Quantum Electrodynamics.

Among various solutions we shall discuss a new one inspired by a proposal of Bethe [14] (in the fermion case) recently discussed by Clark and Pedigo [15]. As we shall show it remains valid despite various objections [16, 17] which have been raised against an initial model [18].

Let us denote by ϕ_γ and ϕ_φ the spinors associated to γ and φ in the de Broglie fusion formalism. They satisfy the relations $\eta^+\phi = \phi_\gamma$ and $\eta^-\phi = \phi_\varphi$ with

$$\left(\beta_\mu \partial^\mu + i\frac{mc}{\hbar}\right)\phi = 0$$

where the β's have 16 dimensions. Both γ and φ are locally endowed with angular momenta $m_{\mu\nu} = (\hbar/2mc)\,\phi^+\beta_\mu\beta_\nu\phi$: $\eta^\pm$ representing projection operators and ϕ the product (fusion) of two spin $\frac{1}{2}$ components.

We then assume that the interaction Hamiltonian between γ and φ takes the form (in de Broglie's notations):

$$H_{\mathrm{I}} = -g_0 : (\bar{\phi}_\varphi (m^{\mu\nu})_\varphi \phi_\varphi)(\bar{U}(m_{\mu\nu})_w W) : \\ -g_i (\bar{\phi}_\gamma (\beta_\alpha)_\gamma \phi_\gamma)(\bar{U}(\beta^\alpha)_w W) : + \text{h.c.}$$

where w stands for an intermediate particle (vector) exchanged between a spin $\frac{1}{2}$ γ and φ component, and $U = w(0)$ for its annihilation state: so that one exchanges two W, i.e. a scalar particle, between γ and φ.

H_{I} yields [19] for the differential cross section the expression

$$\frac{\mathrm{d}\sigma}{\mathrm{d}T} = k \cdot \frac{E-T}{ET}$$

with $T = E - E'$, $E(E')$ representing the initial (final) energy of the incident γ which evidently favors a strong forward scattering for $T \to 0$. The total cross section σ_t is obtained by integrating between T_{min} and T_{max} i.e.

$$\sigma_t = k \log(E/T_{\mathrm{min}} - 1)$$

The average energy loss per collision can be written

$$\langle T \rangle = \frac{1}{\sigma_t} \int T \left(\frac{\mathrm{d}\sigma}{\mathrm{d}T} \right) \mathrm{d}T$$

which yields (per unit of length in a density $\rho_\varphi(r)$):

$$\frac{\mathrm{d}E}{\mathrm{d}r} = -\rho_\varphi(r)\, \sigma_t \left(\frac{\langle T \rangle}{E} \right) E$$

A simple calculation then shows that

$$B = \sigma_t \left(\frac{\langle T \rangle}{E} \right) = \int \left(\frac{T}{E} \right) \left(\frac{\mathrm{d}\sigma}{\mathrm{d}T} \right) \mathrm{d}T = \frac{k}{2} = \text{const.}$$

i.e.

$$\delta z = -\frac{\delta v}{v} = -B \int \varphi_\mu(r)\, \mathrm{d}r$$

in agreement with our initial model [18]. From relativistic kinematics one gets:

$$T = E^2 \cdot \frac{1 - \cos\theta}{2m_\varphi c^2 + E(1 - \cos\theta)} = \frac{2(E^2/m_\varphi c^2)\sin^2(\theta/2)}{1 + 2(E/m_\varphi c^2)\sin^2(\theta/2)}$$

for n collisions we obtain

$$\delta\theta_t = \sqrt{n}\langle\delta\theta\rangle$$

$\langle\delta\theta\rangle$ representing the angular deflection per collision i.e.

$$\frac{\langle T\rangle}{E} > E\frac{\langle\delta\theta\rangle^2}{2m_\varphi c^2}$$

It is clear that the φ particles do not intrude directly into Q.E.D. since they are neutral. Their indirect intrusion via graphs of the type

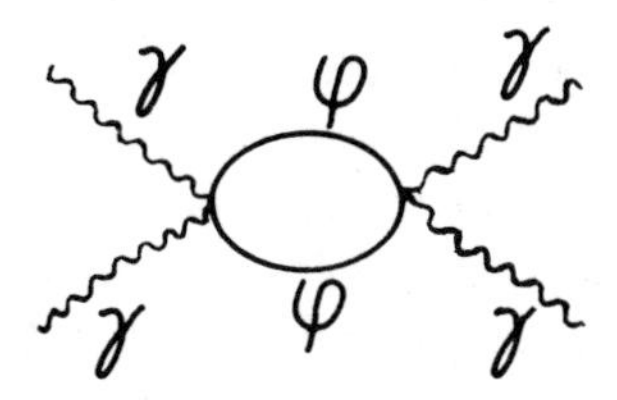

is negligible. The disintegration in flight

$$\gamma + \varphi \to \gamma + \gamma + \cdots + \varphi$$

might explain anomalous phenomena recently discussed by Collins *et al.* if it corresponds to the cross section $\sigma \sim 3 \times 10^{-25}$ cm^2.

The phenomenon discovered by Goldstein [11, 12] helps to fix possible numerical values for H_1.

In the φ-bath surrounding the Sun: if we accept $n_\varphi \sim (1/2)\, n_\gamma$ we find [20] for ~ 100 collisions between the sun and earth $\sigma_t \sim 10^{-22}$ cm^2. This yields a lower value for m_φ since we have $2m_\varphi c^2 \ll h\nu/\delta z$ i.e. (since $\delta\nu/\nu \sim 10^{-9}$) $m_\varphi \ll 10^{-29}$ gr for $\gamma \sim 10^{10}$ Hz.

In conclusion we shall briefly discuss another consequence of the phase correlation principle which goes beyond the scope of the present paper: i.e. the hidden variable theory of photon behaviour.

If we introduce $\mu \neq 0$ we know that true physical field quantities are no longer the $F_{\mu\nu} = \partial_\mu A_\nu - \partial_\nu A_\mu$ but the vector potentials A_μ since the theory is no longer gauge invariant. Boundary conditions should be written directly on A_μ: an idea already suggested by the Bohm-Aharonov effect which (despite the fact it is globally gauge invariant) implies true physical shifts associated with regions where $F_{\mu\nu} = 0$ but $A_\mu = \text{const} \neq 0$. If we introduce complex vector potentials $A_\mu = a_\mu \exp(iS/\hbar)$ where $a_\mu(a_i, v)$ and

S represent real vector and scalar fields in the field equations

$$\partial_\mu F_{\mu\nu} = \mu^2 A_\mu \qquad \partial_\mu \tilde{F}_{\mu\nu} = 0$$

we seen that one can introduce a photon current j_μ and photon spin σ_μ respectively proportional to $(A_\mu \tilde{F}^{\mu\nu *} + cc)$ and $(A_\mu \tilde{F}^{\mu\nu *} + cc)$. The photons in the 'pilot wave' interpretation are then assumed to follow the j_μ lines of flow and are distributed among them by 'quantum jumps' related with subquantum fluctuations [21]. We can explain in this way the quantum probability distribution of photons carried along any given field. This localization of photon yields, as shown by Bell [22] and other authors [23] physically detectable consequences when we analyse various improved versions of the Einstein-Podolsky-Rosen [24] experiment. They are now being tested. Though recent experiments yield conflicting evidence at present it appears that if Holt's and Faraci *et al.*'s [25, 26] results are confirmed, one should consider this as an indirect proof not only of hidden-variable theory but also of the correct nature and depth of de Broglie's original intuition i.e. of the 'phase correlation' principle.

BIBLIOGRAPHY

[1] de Broglie, L., *C.R. Ac. Sci.* **277B** (1973), 71.
[2] Bass, L. and Schrödinger, E., *Proc. Roy. Soc.* **232A** (1955), 1.
[3] Arp, H., *Science* **174** (1971), 1189, and Arp, H., *IAU Symposium* **58**, Report Canberra (1973).
[4] Collins, G. B., Ficenek, J. R., Stevens, D., Trower, W., and Fisher, J., *Phys. Rev.* **982** (1973).
[5] Adler, S. and Bardeen, W., *Phys. Rev.* **D734** (1974).
[6] Feynman, R. F., *Phys. Rev.* **76** (1949), 749, 769.
[7] Souriau, J. M., *Nuovo Cimento* **30** (1963), 565.
[8] Hua, L. K., *Trans. Amer. Math. Soc.* (Providence, R.I., 1963), vol. 6.
[9] Wyler, A., 'The Complex Light-Cone, Symmetric Space of the Conformal Group', Institute for Advanced Study, Preprint, 1972.
[10] Roddier, F., *Ann. Astrophys.* **28** (1965), 478.
[11] Goldstein, R. M., *Science* **166** (1969), 598.
[12] Merat, P., Pecker, J. C., Vigier, J. P., *Astron. Astrophys.* **30** (1974), 167.
[13] Kuhi, L. V., Pecker, J. C., Vigier, J. P., *Astron. Astrophys.* **32** (1974), 111.
[14] Bethe, H. A., *Proc. Comb. Phil. Soc.* **31** (1973), 108.
[15] Clark, R. B. and Pedigo, R. D., *Phys. Rev.* **D**, (1973), 2261.
[16] Aldrovandi *et al.*, *Nature* **241** (1973), 340.
[17] Chew, H., *Nature* **242** (1973), 5.

[18] Pecker, J. C., Roberts, A. P., and Vigier, J. P., *Nature* **241** (1973), 338.
[19] Moles, M. and Vigier, J. P., *C.R. Acad. Sci.* **278 B** (1974), 969.
[20] Ter Haar, D., *Phil. Mag.* **45** (1954), 1023.
[21] Bohm, D. and Vigier, J. P., *Phys. Rev.* **96** (1954), 208.
[22] Bell, J. S., *Physics (Long Is. N.Y.)* **1** (1964), 195.
[23] Clauser, J. F. *et al.*, *Phys. Rev. Lett.* **23** (1969), 880.
[24] Einstein, A., Podolsky, B., and Rosen, N., *Phys. Rev.* **47** (1935), 777.
[25] Holt, R. A., 'Atomic Cascade Experiments', Thesis, Harvard University, 1973.
[26] Faraci, G. *et al.*, Third Int. Conf. on Position Annihilation, Helsinki, 1973; and *Lettere al Nuovo Cimento* **9** (1974), 607.

INDEX OF NAMES